Impossible?

Impossible?

SURPRISING SOLUTIONS TO
COUNTERINTUITIVE
CONUNDRUMS

Julian Havil

PRINCETON UNIVERSITY PRESS
PRINCETON AND OXFORD

Published by Princeton University Press,
41 William Street, Princeton, New Jersey 08540

In the United Kingdom: Princeton University Press,
6 Oxford Street, Woodstock, Oxfordshire OX20 1TW

Library of Congress Cataloging-in-Publication Data
Havil, Julian, 1952–
Impossible? : surprising solutions to
counterintuitive conundrums / Julian Havil.
p. cm.
Includes index.
ISBN 978-0-691-13131-3 (cloth : alk. paper)
1. Mathematics–Miscellanea. 2. Paradox–Mathematics.
3. Problem solving–Miscellanea. I. Title.
QA99.H379 2008
510–dc22   2007051792

British Library Cataloguing-in-Publication Data
A catalogue record for this book is available from the British Library

This book has been composed in LucidaBright
Typeset by T&T Productions Ltd, London
Printed on acid-free paper ∞
press.princeton.edu
Printed in the United States of America

1 3 5 7 9 10 8 6 4 2

In appreciation of
the incomparable Martin Gardner

Martin Gardner has turned dozens of innocent youngsters into math professors and thousands of math professors into innocent youngsters.

Persi Diaconis

Medicine makes people ill, mathematics makes them sad,
and theology makes them sinful.

Martin Luther

I am a mathematician to this extent: I can follow triple
integrals if they are done slowly on a large blackboard by
a personal friend.

J. W. McReynolds

I know that you believe that you understood what you
think I said, but I am not sure you realize that what you
heard is not what I meant.

Robert McCloskey

All truth passes through three stages. First, it is ridiculed.
Second, it is violently opposed. Third, it is accepted as
being self-evident.

Arthur Schopenhauer

First you're an unknown, then you write a book and move
up to obscurity.

Martin Myers

Contents

Acknowledgments

These are old fond paradoxes to make fools laugh i' the alehouse.

Othello, act 1, scene 1

It is a pleasure to acknowledge the understanding and support once again shown by my headmaster, Dr Ralph Townsend, as it is to thank my various classes of young men on whom some of the material has openly or surreptitiously been tested. A special thanks to two of them: Freddie Manners and Miles Wu for some greatly appreciated programming skills.

Thanks to Microsoft and Design Science who, with Word™ and Mathtype™, provided the tools that allowed me to write the many words and symbols very conveniently. Thanks for TEX and to its inventor Donald Knuth for providing the means of transforming the source files into the finished product. Appreciation of Wolfram Research should also be recorded, since almost all graphs and tables have been drawn and constructed by using their product Mathematica™. Also, thanks to Pedagoguery Software for the program Tess™, which was used to produce the motifs. Vickie Kearn, Executive Editor at Princeton University Press, has eased the sometimes challenging book creation process; all thanks to her. Once again, Jon Wainwright of T&T Productions Ltd has shown great expertise and patience in converting the manuscript to professional copy; nothing has ever been too much trouble for him.

The habitués of the Hamilton bar of the Wykeham Arms must be mentioned and some named: Colin, Locky, Vince, Phil, George, Brian, Phil.... And finally there is my extended family, too many to mention exhaustively yet too important to forget entirely. It is headed by my wonderful wife Anne and continues in age order with: Rob, Sophie, Tom, Rachael, Caroline, Simon and Daniel. None of the above will ever read the book but without them it wouldn't have been worth writing it.

Impossible?

Introduction

A paradox,
A most ingenious paradox!
We've quips and quibbles heard in flocks,
But none to beat this paradox!

The Pirates of Penzance

We begin with a classic puzzle.

Imagine a rope just long enough to wrap tightly around the equator of a perfectly spherical Earth. Now imagine that the length of the rope is increased by 1 metre and again wrapped around the Earth, supported in a regular way so that it forms an annulus. What is the size of the gap formed between the Earth and the extended rope?

The vast Earth, the tiny 1 metre—surely the rope will be in effect as tight after its extension as before it? Yet, let us perform a small calculation: in standard notation, if $C = 2\pi r$ for the Earth and the original length of the rope and $C + 1 = 2\pi R$ for the rope lengthened, we require the size of

$$R - r = \frac{C + 1}{2\pi} - \frac{C}{2\pi} = \frac{1}{2\pi}.$$

The Cs have cancelled, leaving $1/2\pi = 0.159\ldots$ m $\approx$ 16 cm as the gap. The Earth could have been replaced by any other planet, an orange or a ping-pong ball and the result would have been the same: a fact which is hard to accept even though the reasoning is irrefutable.

In *Nonplussed!* we gathered together a variety of counterintuitive situations and in this sequel we chronicle eighteen more mathematical phenomena which, if allowed to do so, confound one's reason. The criterion for inclusion has been, as it was in *Nonplussed!*, that, in the recent or distant past, the matter has caused the author surprise. We hope that the reader experiences surprise too, although we must once again accept the subjectivity of the reaction.

Again, probability and statistics show themselves to be fertile ground for the counterintuitive, constituting about half of the material of the book, but the remainder is again eclectic and again meant to be so. Some of the material has an interesting history and where possible we have tried to detail this, and some has far-reaching consequences, extending significantly beyond our treatment here, and we have tried to indicate this too. All-in-all, we have collected together and attempted to explain a medley of the confusing.

The mathematical level varies considerably and we have tried to arrange the chapters so that there is a progression, although this has been particularly difficult to achieve with such a collection of material. Sometimes, in order to include a result at all, it has been necessary to sidestep full mathematical rigour and to be content with what we hope is an illuminating argument, convincing if not fully complete. It is not the sole instance, but perhaps the definitive example of this is the Banach–Tarski paradox, which is utterly profound and about which research papers and books have been written; it lives in the heart of abstract mathematics but to have omitted it completely would have been a greater sin than including it, if somewhat superficially. It is utterly unbelievable. As with all the ideas we introduce, the reader can delve deeper if they desire and we have tried to include appropriate references where possible.

We hope that, as these pages are turned, the reader will be reminded of old favourites or informed of new favourites-in-waiting—and that Simon Newcomb's observation about his tables of logarithms, which we discuss in chapter 16, will not prevail here!

Chapter 1

IT'S COMMON KNOWLEDGE

Mathematics is the science which uses easy words for hard ideas.

Edward Kasner and James Newman

Part of Stevie Nicks's 2003 lyrics of the Fleetwood Mac song 'Everybody Finds Out' might describe the reader's reaction to the subject matter of this first chapter:

I know you don't agree...
Well, I know you don't agree.

The song's title also finds its way into the title of an episode of the NBC sitcom television series *Friends*, 'The One Where Everybody Finds Out', which was first aired in February 1999 and which contains the following dialogue:

Rachel: Phoebe just found out about Monica and Chandler.
Joey: You mean how they're friends and nothing more? [Glares at Rachel]
Rachel: No. Joey, she knows! We were at Ugly Naked Guy's apartment and we saw them doing it through the window. [Joey gasps] Actually, we saw them doing it up against the window.
Phoebe: Okay, so now they know that you know and they don't know that Rachel knows?

Joey: Yes, but y'know what? It doesn't matter who knows
    what. Now, enough of us know that we can just tell
    them that we know! Then all the lying and the secrets
    would finally be over!
Phoebe: Or, we could not tell them we know and have a
    little fun of our own.
Rachel: Wh-what do you mean?

We will consider what brings together pop lyrics, popular television and an important idea which purveys a considerable body
of mathematics and its applications.

Common and Mutual Knowledge

It seems unlikely that the above discourse was meant to plumb
the depths of mathematical logic but it does draw an important distinction between two superficially equivalent concepts:
mutual knowledge and *common knowledge*. We might, for example, suggest that in everyday language it is common knowledge
that the capital city of Australia is Canberra and mean that all
people who know of the country will reasonably be aware of that
fact. As another example, we might say that it is common knowledge that all road users know that a red traffic light means 'stop'
and a green traffic light 'go'. The usage of an ordinary expression
such as 'common knowledge' is fine in normal circumstances but
we are going to deal with a stricter interpretation of the term and
distinguish it from its cousin, mutual knowledge.

Mathematics is given to using ordinary words for technical
purposes: *group, ring, field, rational, transcendental*, etc., each
have their standard dictionary definitions yet each of them
means something entirely different, precise and technical within
the mathematical world. The same is true of the phrase *common knowledge*, the everyday use of which suggests that what
is being referred to is known to all. The crucial point is that it
would not matter very much whether or not an individual knows
that another knows that the capital city of Australia is Canberra,
but in order to ensure safe traffic flow it is not sufficient that

all road users are aware of the colour convention used with traffic lights; it must be the case that they know that all other road users are aware of the convention, otherwise a driver might see a car approaching a red light and wonder whether or not its driver is aware of the convention to stop there.

So, we make two definitions. The first is that of *mutual knowledge*. A statement S is said to be *mutual knowledge* among a group of people if each person in that group knows S. Mutual knowledge by itself implies nothing about what, if any, knowledge anyone attributes to anyone else. It is sufficient that the Canberra example is one of mutual knowledge. The technical definition of *common knowledge* brings about a deeper implication: that everyone knows that everyone knows (and everyone knows that everyone knows that everyone knows, and so on) S. The traffic light example requires common knowledge.

In the dialogue above, Phoebe's statement,

> Okay, so now they know that you know and they don't know that Rachel knows?

distinguishes between the common knowledge shared between Joey and the couple Monica and Chandler and the mutual knowledge shared between Rachel and them.

It is possible to convert mutual into common knowledge. For example, we could assemble a group of strangers in a room and then make the statement: the capital city of Australia is Canberra. If we assume that each individual already knew the fact (and therefore it was already mutual knowledge), at first glance the announcement seems to add nothing, but it has transformed mutual knowledge into common knowledge, with everybody in the room now knowing that everybody in the room knows that the capital city of Australia is Canberra. It is this feature that is central to the main conundrum of the chapter.

There is a well-known example of the phenomenon in children's literature. In Hans Christian Andersen's fable *The Emperor's New Clothes*, two scoundrels convince the vain emperor that they could make a magnificent cloth of silk and gold threads which would be 'invisible to everyone who was stupid or not fit for his post'. After the emperor gave them money and materials to make the royal garments, they dressed him in nothing at all. Not even the emperor, much less his courtiers, dared admit to not seeing any clothes for fear of being branded stupid or incompetent. A ceremonial parade was arranged in order to display the wondrous new clothes and the public applauded as the emperor passed by.

> All the people standing by and at the windows cheered and cried, 'Oh, how splendid are the emperor's new clothes.'

Then a child commented,

> But he hasn't got anything on.

From that moment, what had been the mutual knowledge that the emperor was naked became common knowledge.

This is more than semantic pedantry and we will consider an infamous example of the implication of converting mutual to common knowledge. The technique of mathematical induction will be used and this is reviewed in the appendix (page 221).

A Case of Red and Blue Hats

Suppose that a group of people is assembled in a room and also a number of hats, one for each, coloured either red or blue (accepting that all could be of one colour). For definiteness we will suppose that exactly fifteen of the hats are red, which means that the remainder are blue, although the participants are not aware of this distribution. We will also suppose that each individual is a perfect logician.

A hat is placed on each person's head in such a way that its colour is unknown to that individual but is seen by everyone else. The group of people then sits in the room looking at each other, without communication, and with a clock, which strikes every hour on the hour, available for all to see and hear. Each is instructed to leave the room immediately after the clock strike after which they are certain that they are wearing a red hat.

The group will simply sit in the room, waiting as the clock strikes hour after hour. Those wearing a red hat will see fourteen red hats and those wearing a blue hat will see fifteen red hats; with no extra information, none of them can be certain of the colour of the hat they are wearing: are there fourteen, fifteen or sixteen red hats? Fortunately, a visitor arrives in the room, looks around at the hats being worn and announces, 'At least one person here is wearing a red hat.'

This hardly seems revelatory. Notwithstanding the seeming irrelevance of the announcement, once it is made it is certain that after the subsequent fifteenth strike of the clock, all fifteen people who are wearing red hats will simultaneously walk out of the room.

To consider the reasoning it will be convenient to adopt some notation. Represent the statement 'at least one person is wearing a red hat' by the symbol R_1 and the statement 'A knows X' by the expression $A \rightarrow X$.

First, consider the case of one red hat. Before the statement the wearer, A, sees all blue hats and can have no idea of the colour of his own hat; that is, $A \nrightarrow R_1$. After the announcement $A \rightarrow R_1$ and he will be certain that his hat is red and will walk out after the next clock strike, the first after the announcement. The information that was conveyed by the announcement results in an immediate resolution of the situation.

Now we will deal with two red hats. Before the announcement, R_1 is mutual but not common knowledge. That is, everyone can see at least one red hat and, if the wearers of the red hats are A and B, then $A \rightarrow R_1$ and $B \rightarrow R_1$, since each can see the other's red hat. Yet, $A \nrightarrow (B \rightarrow R_1)$, since $B \rightarrow R_1$ is a direct result of B seeing A's red hat and A has no idea whether or not his hat is indeed red. The announcement tells everybody that R_1 is true and so it is now the case that $A \rightarrow (B \rightarrow R_1)$ (and $B \rightarrow (A \rightarrow R_1)$). Information has been acquired by the announcement; what was mutual knowledge among the red-hat wearers has become common knowledge among them. Now the clock strikes for the first time and none can conclude the colour of their hat: it could be that there is one red hat, in which case, from A's point of view, B is wearing it. Then it strikes a second time and matters change. A argues that, since B did not leave after the first strike of the clock, it must be that he saw a red hat and therefore that there are two such, one on each of the heads of A and B: both will leave the room.

With three red hats, before the announcement the following typifies the situation for red-hat wearers A, B and C: $B \to R_1$, as B can see red hats on both A and C; $A \to (B \to R_1)$, as A can see a red hat on C, but $C \not\to (A \to (B \to R_1))$, since C has no idea whether his hat is red or blue.

After the announcement, R_1 again becomes common knowledge and so $C \to (A \to (B \to R_1))$ and once again information is contained within the seemingly innocent statement and the same argument as above establishes that all three leave after the third strike of the clock.

The reasoning continues with ever deeper levels of knowledge gained as the number of red hats grows with the announcement causing the strike-out of the first arrow in the knowledge chain to disappear in every case: everybody knows that everybody knows that... there is at least one red hat. From this, it is a matter of waiting in order to exclude all possibilities until in the end only one remains; in our case, that all fifteen red head wearers know that there are precisely fifteen red hats.

The 'reasoning continues' type of argument is one which is normally susceptible to proof by induction and we give one such below.

The induction is taken over the clock strike, with R_i taken to mean 'at least i people are wearing red hats'. Now suppose that at the ith strike of the clock R_i is common knowledge. If no red-hatted individual can tell if his hat is red, it must be that R_{i+1} is true since each must be seeing at least i red hats, otherwise he will be able to tell that his hat is red; this together with his own red hat gives the result. This means that on the fifteenth strike of the clock that there are at least fifteen red hats is common knowledge; but the red-hat wearers can only see fourteen red hats and so they must conclude that their hat is indeed red, and will walk out.

The puzzle is one of many variants—with luminaries such as John Edensor Littlewood giving their names to some of them—they all reduce to the same fundamental concept and they are all very, very confusing!

The importance of common knowledge extends far and wide in mathematical application, including the fields of economics, game theory, philosophy, artificial intelligence and psychology. Perhaps the concept dates back as far as 1739 when, in his *Treatise of Human Nature*, the Scottish philosopher David Hume argued that, in order to engage in coordinated activity, all participants must know what behaviour to expect from each other. It is not difficult for the modern author to have empathy with Hume when he (too critically) judged the initial public reaction to the work as such that it 'fell dead-born from the press, without reaching such distinction as even to excite a murmur among the zealots.' It is now generally considered to be one of the most important books in the development of modern philosophy.

As a final problem, we will consider a situation reminiscent of the above in that a striking clock counts out seemingly irrelevant time periods but in which a seemingly irrelevant statement is replaced by a seemingly unhelpful condition.

Consecutive Integers

Two people, A and B, are assigned positive integers; secretly, they are each told their integer and also that the two integers are consecutive. The two sit in a room in which there is a clock, which strikes every hour on the hour. They may not communicate in any way, but they are instructed to wait in the room until one knows the other's number and then to announce that number after the strike of the clock following the revelation of that information.

Both seem destined to stay in the room forever. The clock will relentlessly strike the hour with the two participants seemingly waiting for help that never comes: imagine sitting in the room with, for example, the knowledge that your number is 57; you can have no idea whether the other number is 56 or 58—or can you?

In fact, there is a hidden advantage in the clock striking and knowing that the numbers are consecutive, which our intuition can easily fail to exploit. A careful use of induction can succeed in that exploitation, and having done so should convince us that at some stage one of the two people will leave the room.

To get a feel for what is really happening, suppose that A's number is 1, then it must be that B's number is 2 and after the first strike of the clock A will announce that B has the number 2. Now take the next case and suppose that A's number is 2. This means that B's number is either 1 or 3. If it is 1, B will announce after the first strike of the clock, as above; if the announcement is not made, A will know that B's number is 3 and announce this fact after the second strike of the clock. The argument can be continued methodically and is best done so using induction to give the remarkable result that *the person whose number is n will announce that the other player's number is n + 1 after the nth strike of the clock.*

In fact, the proof is easy. We have already argued that the statement is true if the lower number is 1. Now let us suppose that the statement is true when the lower number is k and that A is given the number $k + 1$. Then if B holds k, by the induction hypothesis he will announce A's number after the kth strike of the clock, otherwise B holds $k + 2$ and A will know this to be the case after the kth strike of the clock and so announce B's number after the $(k + 1)$th strike of the clock, and the induction is complete.

Chapter 2

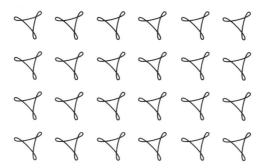

SIMPSON'S PARADOX

Statistics are like bikinis. What they reveal is suggestive, but what they conceal is vital.

Aaron Levenstein

Apocryphal Stories

It is difficult for the non-cricket fanatic to appreciate the trauma associated with the biannual cricket competition between the arch-rivals England and Australia, universally known as the *Ashes*. On 29 August 1882 (at home) a full-strength England cricket side was for the first time beaten by Australia, which caused the British publication *The Sporting Times* to run an obituary for English cricket which included the words 'The body will be cremated and the Ashes taken to Australia'. On the return fixture (in Australia) England regained the upper hand and a small urn was presented to the captain, Lord Darnley, in commemoration; and so the uncompromisingly fierce competition began for the notional possession of a tiny urn of questionable contents which hardly ever leaves London no matter who wins it.

A chance hit on a Queensland educational website[1] revealed a little apocryphal story based on the two former Australian batsmen, the brothers Steve and Mark Waugh. In paraphrase it read

> Steve and Mark decided to have a little wager on who would have the better overall batting average over the two upcoming Ashes series, the first in England and the second in Australia.
>
> After the first Ashes series, Steve said to Mark, 'You've got your work cut out for you, mate. I have scored 500 runs for 10 outs, for an average of 50. You have 270 runs for 6 outs, for an average of 45.'
>
> After the second Ashes series, Steve continued by saying, 'Ok, mate, pay up. In this series I scored 320 runs for 4 outs, an average of 80, while you had 700 runs for 10 outs, which is only an average of 70. I topped you in each of the Series.'
>
> 'Hold on,' said Mark, 'The wager was for the better batting average overall, not series by series. As I reckon it, you have scored 820 runs for 14 outs, and I have scored 970 runs for 16 outs. Your average is 58.6, while my average is 60.6. I win.'

How is this possible, that Steve could have the better average in each of the two tests but a lower average overall?

The matter at hand has nothing to do with the intricacies of cricket. The *Ask Marilyn* column in *Parade Magazine* (a supplement to many American Sunday newspapers) provides a forum for readers to ask questions and give opinions on a wide variety of matters and often generates a great deal of reader response. Sometimes readers send in questions for the column's editor, Marilyn Vos Savant, to contemplate—and since she is listed in the *Guinness Book of World Records Hall of Fame* as the individual with the highest IQ, they can reasonably expect thought-provoking answers. The following question was posed by a reader in the *Ask Marilyn* column in the 28 April 1996 issue of *Parade Magazine*:

---

[1] http://exploringdata.cqu.edu.au/sim_par.htm.

Table 2.1.

	Population		Deaths		Death rate per 100 000	
	NY	R'd	NY	R'd	NY	R'd
Caucasians	4 675 174	80 895	8365	131	179	162
African Americans	91 709	46 733	513	155	560	332
Totals	4 766 833	127 268	8881	286	187	226

A company decided to expand, so it opened a factory generating 455 jobs. For the 70 white collar positions, 200 males and 200 females applied. Of the females who applied, 20% were hired, while only 15% of the males were hired. Of the 400 males applying for the blue collar positions, 75% were hired, while 85% of the 100 females who applied were hired.

A federal Equal Employment enforcement official noted that many more males were hired than females, and decided to investigate. Responding to charges of irregularities in hiring, the company president denied any discrimination, pointing out that in both the white collar and blue collar fields, the percentage of female applicants hired was greater than it was for males.

But the government official produced his own statistics, which showed that a female applying for a job had a 58% chance of being denied employment while male applicants had only a 45% denial rate. As the current law is written, this constituted a violation. Can you explain how two opposing statistical outcomes are reached from the same raw data?

The reader may wish to check the arithmetic but Marilyn correctly noted that, even though all the figures presented are correct, the two outcomes are not, in fact, opposing. Nor is it the case that such conflicting data are necessarily contrived. Consider the following true story.

In 1934 Morris Cohen and Ernst Nagel cited actual 1910 death rates from tuberculosis in two cities (Richmond, Virginia, and New York, New York). Table 2.1 shows their data. From it we

can see that the death rates for Caucasians and African Americans were each individually lower in Richmond than in New York,
yet the death rate for the total combined population of African
Americans and Caucasians was higher in Richmond than in New
York.

Simpson's Paradox

All of the above are examples of sets of data separately support-
ing a certain hypothesis but, when combined, support the opposite hypothesis. The phenomenon is known as *Simpson's Paradox*, after E. H. Simpson, who discussed it in a 1951 article (The
interpretation of interaction in contingency tables, *Journal of the
Royal Statistical Society* B 13:238–41). As is so often the case, the
person after whom a result is named is not the person who first
considered it. G. Udny Yule preceded Simpson in 1903 (Notes on
the theory of association of attributes in statistics, *Biometrika*
2:121–34) and he was preceded by Karl Pearson, A. Lee and L.
Bramley-Moore in 1899 (Genetic (reproductive) selection: inheri-
tance of fertility in man, *Philosophical Transactions of the Royal
Statistical Society* A 173:534–39): Yule described the association
as '*spurious*' or '*illusory*'. Yet, it was Simpson's witty and surprising illustrations of the phenomenon which earned the name
and the clear view that something peculiar but explicable was
happening.

As our contender for a witty illustration, consider the following factual case, which demonstrates the process in reverse.

An argument to substantiate the claim that foreigners were
more likely to be insane than native-born Americans was advanced in Massachusetts in 1854 and table 2.2 shows the figures
that were given in justification. These show that the probability
that a foreign-born individual was deemed insane was $\frac{625}{230\,000} =
2.7 \times 10^{-3}$, whereas for a native-born individual the probability
reduces to $\frac{2007}{894\,676} = 2.2 \times 10^{-3}$. There might be something in the
claim.

Now let us agree to divide the data according to an accepted
social hierarchy of the time: rather strange to the modern eye

Table 2.2. Whole population.

	Insane	Not insane	Totals
Foreign born	625	229 375	230 000
Native born	2007	892 669	894 676
Totals	2632	1 122 044	1 124 676

Table 2.3. Pauper class.

	Insane	Not insane	Totals
Foreign born	182	9 090	9 272
Native born	250	12 513	12 763
Totals	432	21 603	22 035

Table 2.4. Independent class.

	Insane	Not insane	Totals
Foreign born	443	220 285	220 728
Native born	1757	880 156	881 913
Totals	2200	1 100 441	1 102 641

the division is into the *pauper* class and the *independent* class. We then arrive at tables 2.3 and 2.4.

Within the pauper class we have that the probability of a foreign-born person being deemed insane is $\frac{182}{9272} = 0.02$, which is the same as a native-born person, with the calculation $\frac{250}{12763} = 0.02$. The same is true for the independent class, where the probabilities are both 2.0×10^{-3}; so, if an adjustment is made for the status of the individuals, we see that there is no relationship at all between sanity and origin.

An Analysis

For the purposes of illustration we will detail a final, theoretical example of the phenomenon.

Table 2.5. Effects of the drugs on men.

	C	$\sim C$	
X	40	160	200
Y	30	170	200
	70	330	400

Table 2.6. Effects of the drugs on women.

	C	$\sim C$	
X	85	15	100
Y	300	100	400
	385	115	500

Suppose that two new drugs, X and Y, are tested on a sample of the population suffering from a particular ailment and that tables 2.5 and 2.6 show the comparison of the effectiveness of the two drugs on men and women separately, giving frequencies of curing the patient (C) and otherwise ($\sim C$). Since $\frac{40}{200} > \frac{30}{200}$ and $\frac{85}{100} > \frac{300}{400}$ the tables show that for both males and females drug X is more effective than drug Y.

Now combine the data to arrive at table 2.7, which shows the comparative effect of each drug for the population as a whole. Since $\frac{330}{600} > \frac{125}{300}$, drug Y is now more effective than drug X. Which is better, drug X or drug Y?

The structure of the process is encapsulated in tables 2.8–2.10.

We can see that the basis for the paradox is the simple arithmetical fact that, for positive numbers, if

$$\frac{a}{b} > \frac{c}{d} \quad \text{and} \quad \frac{p}{q} > \frac{r}{s},$$

it is not necessarily the case that

$$\frac{a+p}{b+q} > \frac{c+r}{d+s},$$

and vice versa. For example, $\frac{1}{2} > \frac{3}{7}$ and $\frac{1}{5} > \frac{1}{6}$ but

$$\frac{1+1}{2+5} = \frac{2}{7} < \frac{4}{13} = \frac{3+1}{7+6}.$$

Table 2.7. Effects of the drugs on both sexes combined.

	C	$\sim C$	
X	125	175	300
Y	330	270	600
	455	445	900

Table 2.8. Subcollection 1.

	Possess attribute	Do not possess attribute	Totals
Attribute	a	$b - a$	b
Alternative attribute	c	$d - c$	d
Totals	$a + c$	$b - a + d - c$	$b + d$

Table 2.9. Subcollection 2.

	Possess attribute	Do not possess attribute	Totals
Attribute	p	$q - p$	q
Alternative attribute	r	$s - r$	s
Totals	$p + r$	$q - p + s - r$	$q + s$

Table 2.10. Total sample.

	Possess attribute	Do not possess attribute	Totals
Attribute	$a + p$	$b - a + q - p$	$b + q$
Alternative attribute	$c + r$	$d - c + s - r$	$d + s$
Totals	$a + c + p + r$	$q - p + s - r + b - a + d - c$	$b + d + q + s$

In the hypothetical case above, $\frac{40}{200} > \frac{30}{200}$ and $\frac{85}{100} > \frac{300}{400}$ but

$$\frac{40+30}{200+200} = \frac{70}{400} < \frac{385}{500} = \frac{85+300}{100+400}.$$

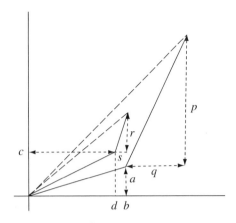

Figure 2.1.

If we use matrix notation to summarize the data sets with

$$X = \begin{pmatrix} a & b \\ c & d \end{pmatrix} \quad \text{and} \quad Y = \begin{pmatrix} p & q \\ r & s \end{pmatrix}$$

representing the two subcollections, then

$$X + Y = \begin{pmatrix} a & b \\ c & d \end{pmatrix} + \begin{pmatrix} p & q \\ r & s \end{pmatrix} = \begin{pmatrix} a+p & b+q \\ c+r & d+s \end{pmatrix},$$

in which case the above conditions on inequalities translate to the equally obvious statement that, if $\det X > 0$ and $\det Y > 0$, it is not necessarily the case that $\det(X + Y) > 0$: matrix determinants are not additive.

Alternatively, we can use figure 2.1 to provide a geometric explanation of how the reversal can occur, taking lines the slopes of which are the fractions we are comparing: the slopes of the dashed lines are in the reverse order to the slopes of the comparable full lines.

As an aid to constructing paradoxical data we can argue as follows.

Since we assume that

$$\frac{p}{q} > \frac{r}{s},$$

we have

$$p > q\frac{r}{s},$$

and if the paradox is to exist we reverse the third inequality to get

$$\frac{a + p}{b + q} < \frac{c + r}{d + s}$$

and so

$$p < \frac{c + r}{d + s}(b + q) - a.$$

These combine to bounds on p of

$$q\frac{r}{s} < p < \frac{c + r}{d + s}(b + q) - a.$$

For example, we can determine the boundaries for p which allow the paradox to exist using the data from the theoretical example to get $75 < p < 125$, and the actual value of $p = 85$ sits nicely in that interval.

To find a lower bound on q we can eliminate p above to give

$$q\frac{r}{s} < \frac{c + r}{d + s}(b + q) - a$$

and therefore we have

$$q\left(\frac{r}{s} - \frac{c + r}{d + s}\right) < b\frac{c + r}{d + s} - a$$

or

$$q(r(d + s) - s(c + r)) < s(b(c + r) - a(d + s)),$$

which means that $q(dr - cs) < s(b(c + r) - a(d + s))$.
If, for the sake of definiteness, we take

$$\frac{r}{s} > \frac{c}{d},$$

and so $dr - cs > 0$, we can now divide to get our inequality

$$q < \frac{s(b(c + r) - a(d + s))}{dr - cs}.$$

Again, from the theoretical example we get $q < 285.7$ and $q = 100$ again fits the inequality.

Finally, it might be of interest to consider the smallest population in which the paradox can exist. If we assume that no category can have a zero entry, then Thomas Bending has shown that

$$a = 1, \quad b = 2, \quad c = 3, \quad d = 7,$$
$$p = 1, \quad q = 5, \quad r = 1, \quad s = 6$$

gives a paradoxical situation with a total population of $b + d + q + s = 20$; whether or not this is minimal, as he says, is quite another matter.

Examples of the occurrence of Simpson's Paradox are legion in areas ranging from SAT scores divided into ethnic groups (D. Berliner, 1993, *Educational Reform in an Era of Disinformation. Educational Policy Analysis Archives*), through growth of children in South Africa (Christopher H. Morrell, 1999, Simpson's Paradox: an example from a longitudinal study in South Africa, *Journal of Statistics Education*, volume 7(3)) to the much publicized Berkeley sex bias case of 1973 in which the University of California at Berkeley was sued for bias against women applying to graduate school (P. J. Bickel, E. A. Hammel and J. W. O'Connell, 1975, Sex bias in graduate admissions: data from Berkeley, *Science* 187:398–404).

Chapter 3

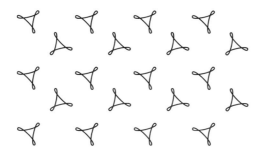

THE IMPOSSIBLE PROBLEM

If you think it's simple, then you have misunderstood the problem.

Bjarne Strustrup

Double Dutch

The Dutch mathematician, mathematical historian and educator Hans Freudenthal was an original and inspirational thinker. Radio telescopes are pretty complicated mechanisms. Freudenthal therefore (and reasonably) argued that electronic communication with extraterrestrial life would require of them the capacity to count and to recognize that $2 + 2 = 4$ and from this conviction he created a mathematically based interstellar language called Lingua Cosmica (the *language of the cosmos*, which was published in his book *LINCOS: Design of a Language for Cosmic Intercourse* in 1960). It seems that he also created a remarkable logical puzzle, which we will consider in this chapter. It was published in 1969 as problem number 223 in the Dutch *Nieuw*

Archief voor Wiskunde (*New Archive for Mathematics*), and in its original form looked like:

No. 223. *A* zegt tot *S* en *P*: Ik heb twee gehele getallen *x*, *y* gekozen met $1 < x < y$ en $x + y \leqslant 100$. Straks deel ik $s = x + y$ aan *S* alleen mee, en $p = xy$ aan *P* alleen. Deze mededelingen blijven geheim. Maar jullie moeten je inspannen om het paar $\{x, y\}$ uit te rekenen.

Hij doet zoals aangekondigd. Nu volgt dit gesprek:

1. *P* zegt: Ik weet het niet.
2. *S* zegt: Dat wist ik al.
3. *P* segt: Nu weet ik het.
4. *S* zegt: Nu weet ik het ook.

Bepaal het paar $\{x, y\}$. (H. Freudenthal)

The reader may appreciate the following translation:

No. 223. *A* says to *S* and *P*: I have chosen two integers *x*, *y* such that $1 < x < y$ and $x + y \leqslant 100$. In a moment I will inform *S* only of $s = x + y$, and *P* only of $p = xy$. These announcements will be private. You are required to determine the pair $\{x, y\}$. He acts as promised. Then the following conversation takes place:

1. *P* says: I do not know the pair.
2. *S* says: I knew you didn't.
3. *P* says: I now know it.
4. *S* says: I know it too.

Determine the pair $\{x, y\}$. (H. Freudenthal)

Other and Later Versions

The puzzle in its English form above reappeared as problem 977 in the *Problem* section of the March 1976 issue of *Mathematics Magazine* (volume 49(2)), submitted by David J. Sprows. An editor's footnote describes it as

A succinct variation of some past problems in the *American Mathematical Monthly*, especially E776, E1126 and E1156,

an observation which dates the problem's variations at 1948, 1955 and 1956 respectively, and assuredly they are significant variations, which as we will see is a highly noteworthy matter. It is almost inevitable that Martin Gardner was responsible for bringing the problem to the greater mathematical puzzling public and certainly to the attention of this author. In the *Mathematical Games* section of the December 1979 issue of *Scientific American*, which he subtitled, 'A pride of problems, including one that is virtually impossible', Gardner listed several short, unrelated problems to challenge his readers, the first of which is his version of Freudenthal's puzzle, told to him by the late Canadian puzzle and magic genius Mel Stover. The statement was preceded by the opening paragraph:

> It is hoped that the following unrelated problems will be new and challenging to most readers. Number 1 is so difficult, with a solution that would take up an inordinate amount of space next month that it is answered at the end of the column. Readers who relish a tough challenge are urged to work on the problem before they read the solution. If there is a simpler solution to the problem than the one given, I should like to know about it. The other problems will be answered at the end of next month's column.

He termed number 1 'The impossible problem'.

Gardner's original version is as follows:

> Two numbers (not necessarily different) are chosen from the range of positive integers greater than 1 and not greater than 20. Only the sum of the two numbers is given to mathematician S. Only the product of the two is given to mathematician P.
>
> On the telephone S says to P, 'I see no way you can determine my sum.'
>
> An hour later P calls him back to say, 'I know your sum.'
> Later S calls P again to report, 'Now I know your product.'
> What are the two numbers?

He continued by remarking:

> To simplify the problem, I have given it here with an upper
> bound of 20 for each of the two numbers. This means that
> the sum cannot be greater than 40 or the product greater
> than 400. If you succeed in finding the unique solution, you
> will see how easily the problem can be extended by raising
> the upper bound. Surprisingly, if the bound is raised to 100,
> the answer remains the same.

We can see that Gardner's version is a variant of the original,
most particularly and significantly because of the number and
order of the statements. Over the years the problem has spawned
any number of other variants each of which continues to maintain an air of mystery and surprise; it simply does not seem possible to solve any of its forms with the information given, but
the solutions do exist, and involve the use of one of the great
conjectures of mathematics, the Goldbach Conjecture, which is
described in the appendix (page 224) (see Torsten Sillke's page
www.mathematik.uni-bielefeld.de/~sillke/).

The exact wording of any variant is critical. Lee Sallows considered Gardner's version in great detail in his article 'The impossible problem' (1995, *The Mathematical Intelligencer* 17(1):27–33)
and we will look at a particular formulation which has, for reasons of the names' first letters, been framed in terms of the two
perfect logicians Polly and Sam. Notice the upper bound.

> Polly and Sam are visited by a friend. The friend, hav-
> ing thought of two integers between 2 and 800 inclusive,
> whispers their product to Polly and their sum to Sam. The
> following dialogue results:
>   1. Polly: I don't know the two numbers.
>   2. Sam: I know that and neither do I.
>   3. Polly: I know the two numbers.
>   4. Sam: So do I.
> What are the two numbers?

We will draw inferences from the information contained within
Polly's and Sam's statements, but it should be made clear that,

in all likelihood, other inferences than those which we make can be made, which could lead to the elimination process behaving in a different manner to the one described below. This is not so much a puzzle to solve, but one to investigate.

First, although the emphasis it provides is useful, the 'and neither do I' part of Sam's response is redundant. The only way Sam could know the numbers at this stage is if they are the pair $(2, 3)$ and this possibility has already been eliminated by Polly.

Keeping track of the consequences of the deductive processes is greatly assisted by the use of a computer and the problem has long been used as an opportunity for Artificial Intelligence programming, with many programs having been written in languages such as Lisp and Prologue. We will give a general schema in a pseudo-code, which can be translated into a specific language.

An Analysis

If we call the two integers x and y, we can make the following deductions:

After statement 1.

x and y cannot both be prime. If they were, the given product could be factored in only one way and Polly would know the numbers, which would contradict her first statement.

$x \times y$ cannot be the cube of a prime p otherwise Polly would know that the numbers are p and p^2.

After statement 2.

$x + y$ must be odd. This is where we need the Goldbach Conjecture. If $x + y$ is even, using the conjecture, it is possible for it to be written as the sum of two primes. If this were the case, it would again mean that both x and y would be prime, in which case Sam could not be certain that Polly could not deduce the values of x and y.

$x + y$ is not 2 more than a prime. If it were, then x could be 2 and y could be prime, in which case the product would be $2y$ and again Sam could not be certain that Polly could not deduce the values of x and y.

$x + y < 403$. If not, $x + y \geqslant 403$ and this means that x (say) could be the prime 401 with $y \in \{2, 4, 6, \ldots, 800\}$ even, since $x + y$ is odd. This means that the product known to Polly would be $401y$ with the smallest factor of y being 2. It must then be that $x = 401$, otherwise $x \geqslant 2 \times 401 = 802$ and that is out of the allowed range. Therefore, if Sam were in possession of the sum $x + y \geqslant 403$, again he could not be certain that Polly could not deduce the values of x and y.

Using this information a list of allowed number pairs can be formed; call it L.

After statement 3.

Since Polly tells Sam that she can now deduce the two numbers, it must be that her product appears uniquely in the products of the members of L. Polly can look at L, form the products and identify the unique pair that generates her given product.

After statement 4.

Since Sam tells Polly that he can now deduce the two numbers, it must be that his sum is uniquely formed from L. He can do for the sum what Polly did for the product.

With our analysis in place we can develop our pseudo-code and detail the results of a computer program based on it. Before we do this we can reduce the size of the list from its original $799^2 = 638\,401$ entries by assuming that $x \leqslant y$, since addition and multiplication are commutative operations.

Pseudo-Code

Array $A := \{(x, y) : \{x, 2, 800\}, \{y, x, 800\}\}$
$A := \{(2, 2), (2, 3), (2, 4), \ldots, (799, 799), (799, 800), (800, 800)\}$
LENGTH$[A] = 319, 600.$

After statement 1.
$A \rightarrow$ SELECT$[A] : \{(x, y) : (\text{NOT}[\text{PRIME}[x]\&\text{PRIME}[y]])$
$$\&(x \times y \neq \text{prime}^3)\}$$

$A := \{(2,6), (2,8), (2,9), (2,10), (2,12), \ldots,$
$$(798,800), (799,800), (800,800)\}$$
LENGTH$[A]$ = 309, 861.

After statement 2.
$A \to$ SELECT$[A] : \{(x,y) : (x + y \text{ ODD})\&(x + y < 403)$
$$\&(x + y \neq \text{prime} + 2)\}$$
$A := \{(2,9), (2,15), (2,21), (2,25), \ldots,$
$$(198,199), (198,203), (199,202)\}, (200,201)$$
LENGTH$[A]$ = 12, 996.

This is list L.

After statement 3.
$A \to$ SELECT$[A] : \{(x,y) : x \times y \text{ is unique}\}$
$A := \{(2,9), (2,25), (2,27), (2,49), \ldots,$
$$(198,199), (198,203), (199,202), (200,201)\}$$
LENGTH$[A]$ =4,471

Polly searches L for her known product, which must be unique.

After statement 4.
$A \to$ SELECT$[A] : \{(x,y) : x + y \text{ is unique}\}$
$A := \{(4,13)\}$
LENGTH$[A]$ = 1.

Sam searches L for his sum.
    The unique solution is the pair $(4,13)$.
    With a mathematical programming language available, there are functions which will greatly help with this sifting; otherwise some judicious programming is needed!

Further Thoughts

There is significance in there being just one member in the final list in the pseudo-code. It not only means that the provider of the numbers has no choice in them if the dialogue is to be correct, but also that an observer who has listened to Polly and Sam's conversation could also identify the two numbers. To understand the problem completely, it is important to distinguish between what Polly and Sam know and what an observer knows from listening to them. The first two statements make no use of the specific numbers that Polly and Sam have been given and the listening observer could make the same deductions and arrive at the list L. L has many number pairs and Polly uses her knowledge of her number to pick the only pair in it that will do. Sam then does the same with 'product' replaced by 'sum'. In fact, now neither has need of the precise number that was given them as each could both add and multiply the number pairs and arrive at the unique $(4, 13)$. Since this is true the observer could have solved the problem himself and told Polly and Sam what the two numbers are.

We have given the upper bound as 800; now let us vary it. There is an important event when the upper bound is 123. The pair $(4, 61)$ makes a first appearance in L, since it just passes through the sieve made by the first two statements. Having made its appearance, $(4, 61)$ continues to be eliminated, with $4 + 61 = 65$ not unique—until the upper bound is 867—when the pairs that sum to 65 and which cause $(4, 61)$ to be eliminated are themselves eliminated, leaving that pair to pass through, and for the first time. The final list scrutinized by Polly and Sam is then $\{(4, 13), (4, 61)\}$. Now Polly and Sam have to use their knowledge of the exact value to arrive at the answer. For the first time, the provider of the numbers could have chosen a second number pair and the observer is not able to solve the problem.

Notice that $(4, 61)$ cannot be eliminated by Polly since the only possible other pair multiplying to 244 is $(2, 122)$ which has both entries even.

This is no more than a special case of the form $(2^n, p)$, where p is prime, with $(4, 13)$ the first example of it. The product of the two numbers is $2^n p$, which cannot be duplicated in L since to do so would mean moving a 2 across to the prime, and again this makes both numbers even. These numbers can only be eliminated from L by Sam. If other number pairs in L sum to $2^n + p$, the pair will be eliminated, otherwise Sam will have to use his knowledge of the exact value of the sum to make his second statement. In this sense pairs of numbers of this type are the most difficult to eliminate from L and as we increase the upper bound we see more and more of them appear, but to do this we need to generalize the third conclusion we made from statement 2. The earlier argument resulting in the bound of 443 is a special case of the following: if x and y each have an upper bound of n, then $x + y < \lceil\frac{1}{2}n\rceil + 2$, where $\overline{N}$ is the smallest prime greater than or equal to N.

The justification is really the same as before. If $x + y \geq \lceil\frac{1}{2}n\rceil + 2$, one of x or y (say x) might be $\lceil\frac{1}{2}n\rceil$ and y must be an even integer 2 or greater. Polly would then be in possession of the product $y\lceil\frac{1}{2}n\rceil$. She would therefore know the two numbers, since the only possible ambiguity would be in moving a factor from y to $\lceil\frac{1}{2}n\rceil$ and the smallest factor that y can have is 2 and that would make $x = 2 \times \lceil\frac{1}{2}n\rceil > n$.

If we run the program for the upper bound of 2000, the final list is

$$\{\{4, 13\}, \{4, 61\}, \{16, 73\}, \{32, 131\}\},$$

and if we go as far as an upper bound of 5000, it becomes

$$\{\{4, 13\}, \{4, 61\}, \{4, 229\}, \{16, 73\}, \{16, 111\},$$
$$\{32, 131\}, \{32, 311\}, \{64, 73\}, \{64, 309\}, \{67, 82\}\}.$$

And this brings us to the end of our own investigation into the problem!

Three Variants

Finally, we offer the reader three standard but lesser-known variants—but without answers.

Variant 1. Three people V, C and X are joined by another person M, who holds hidden the sixteen cards: A, Q, 4 (♥); J, 8,7, 4, 3, 2 (♠); K, Q, 6, 5, 4 (♣); A, 5 (♦).

M selects a card at random and tells V the card's value and C its colour. After this, in X's hearing, he asks them the question, 'Do you know which card I have?' The following conversation ensues:

V : I don't know what the card is.
C : I knew that you didn't know.
V : I know the card now.
C : I know it too.

X thinks for a moment, and concludes correctly what M's card is. How is this possible?

Variant 2.   Each of A, B and C is wearing a hat on which a positive number is printed. Each can see the numbers on the others' hats but not their own number. All are told that one of the numbers is the sum of the other two. The following statements are made in the hearing of all:

A : I cannot deduce what my number is.
B : I cannot deduce what my number is.
C : I cannot deduce what my number is.
A : I can deduce that my number is 50.

What are the numbers on the other two hats?

Variant 3.   A person M joins two others, A and B. M whispers to A the sum of two natural numbers and to B the sum of the squares of the same two natural numbers. Each knows the nature but not the detail of the information being conveyed. The following conversation takes place:

B : I do not know the numbers.
A : I do not know the numbers.
B : I do not know the numbers.
A : I do not know the numbers.
B : I do not know the numbers.
A : I do not know the numbers.
B : I know the numbers.

What are the two natural numbers?

Chapter 4

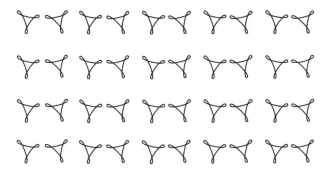

BRAESS'S PARADOX

People take different roads seeking fulfillment and happiness. Just because they're not on your road doesn't mean they've gotten lost.

H. Jackson Browne

A Road to Nowhere

In the 1969 publication 'Graphentheoretische Methoden und ihre Anwendungen', W. Knödel remarked that

> ...major road investment in Stuttgart's city centre, in the vicinity of the Schlossplatz, failed to yield the benefits expected. They were only obtained when a cross street, the lower part of Königstrasse, was subsequently withdrawn from traffic use....

Eliminating a road, rather than building a road, improved traffic flow.

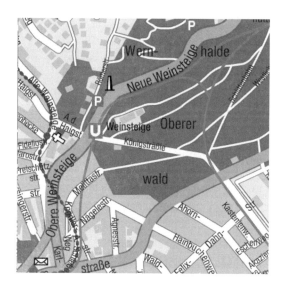

Figure 4.1.

Figure 4.1 shows a map of the relevant part of the city, with part of Königstrasse now a pedestrian precinct.

When 42nd Street in New York City was temporarily closed to traffic, rather than the expected gridlock resulting, traffic flowed more easily; in fact, it was reported in the 2 September 2002 edition of *The New Yorker* that in the twenty-three American cities that added the most new roads per person during the 1990s, traffic congestion rose by more than 70%.

These observed phenomena would have been no surprise to the German mathematician Dietrich Braess, who had published the article 'Über ein Paradoxon aus der Verkehrsplanung' in *Unternehmenstorchung* (12:258–68) in 1968, in which he made exactly that point: under the appropriate conditions, building new roads to ease congestion actually makes the problem worse. He made the point by use of a simple, hypothetical road network which deteriorated when a bypass link was added to it.

The Grip of an Invisible Hand

Figure 4.2 shows Braess's hypothetical system of one-way roads from A to B, via alternatives X and Y. The arrows indicate the direction in which travel is possible and the expressions labelling

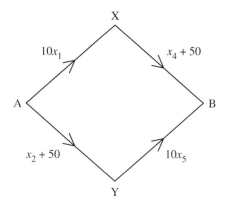

Figure 4.2.

the four routes are the 'cost functions' for that route. Here we can think of the 'cost' of following a particular route as the time taken for it to be traversed, given that a certain number of vehicles are using it; the driver will, we assume, wish to minimize this and the assumption is that he or she will be completely aware of the traffic situation at all times, selfishly choosing the route that is most beneficial to that individual. The determination of such cost functions is a complicated example of mathematical modelling but Braess suggested the simplest model of all: a linear function of the number of vehicles x_r using a particular route, which might be thought of as a loading influenced by the nature of the road in question and which uses up time, fuel, etc.

If we take the traffic load from A to B to be n vehicles (perhaps an hour), the costs of using route X and route Y are clearly $10x_1 + x_4 + 50$ and $10x_5 + x_2 + 50$, respectively, and since any vehicle which committed to using the first part of a route must use the second part, $x_1 = x_4$ and $x_2 = x_5$. This means that the cost functions simplify to $11x_1 + 50$ and $11x_2 + 50$, respectively. The knowledgeable drivers will decide to use a particular route depending on the relative sizes of x_1 and x_2. The first driver in the system will have a choice of either route, the second will switch to the other route, the third is in the same position as the first, etc.: equilibrium is reached with the n drivers split into two sets of $\frac{1}{2}n$, with equal cost functions of $C = 11 \times \frac{1}{2}n + 50 = \frac{1}{2}(11n + 100)$.

With this entirely selfish model in place, the natural equilibrium position is reached with the load equally shared between the two alternatives.

Now let us suppose that the authorities intervene by introducing a mechanism which routes traffic so that the average cost function is minimized, thereby bringing about a collective rather than individual benefit.

With the distribution shown in figure 4.2, the average cost function is given by

$$A = \frac{1}{n}(x_1(11x_1 + 50) + x_2(11x_2 + 50))$$
$$= \frac{1}{n}(11(x_1^2 + x_2^2) + 50(x_1 + x_2))$$
$$= \frac{1}{n}(11(x_1 + x_2)^2 + 50(x_1 + x_2) - 22x_1x_2)$$
$$= \frac{1}{n}(11n^2 + 50n - 22x_1(n - x_1)),$$

since $x_1 + x_2 = n$.

A little extra algebra allows us to rewrite the expression as

$$A = \frac{1}{n}(\tfrac{11}{2}n^2 + 50n + 22(x_1 - \tfrac{1}{2}n)^2)$$

and this is clearly minimized when $x_1 = \frac{1}{2}n$ and so, once again, we have the result that an equally shared load provides the solution to the problem. The selfish approach has done as much as one which possesses corporate responsibility, with individual optimization aggregating to collective optimization: Adam Smith's invisible hand has taken its grip.[1] Yet, Smith's invisible

---

[1] In his 1776 book, *The Wealth of Nations*, the economist Adam Smith wrote:

Every individual necessarily labours to render the annual revenue of the society as great as he can. He generally indeed neither intends to promote the public interest, nor knows how much he is promoting it. He intends only his own gain, and he is in this, as in many other cases, led by an invisible hand to promote an end which was no part of his intention. By pursuing his own interest he frequently promotes that of the society more effectually than when he really intends to promote it. I have never known much good done by those who affected to trade for the public good.

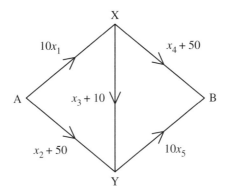

Figure 4.3.

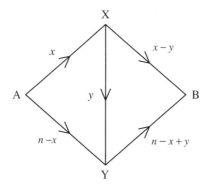

Figure 4.4.

hand sometimes loses its grip—and we are about to approach one such example.

The Grip Loosens

Suppose that, with a view to easing congestion, a one-way relief road is built joining X and Y, with a cost function of $x_3 + 10$, as shown in figure 4.3.

We can no longer conclude that $x_1 = x_4$ and $x_2 = x_5$ and we will study the behaviour of the network by adopting the notation that, of the n vehicles entering A, x of them choose path AX and y of those subsequently choose path XY. This means that $x_1 = x$, $x_2 = n - x$, $x_3 = y$, $x_4 = x - y$ and $x_5 = n - x + y$, as in figure 4.4.

Now we can compute the cost functions for each of the three routes from A to B via X or via Y or via X and Y:

$$C_X = 10x + (x - y) + 50$$
$$= 11x - y + 50,$$
$$C_Y = (n - x) + 50 + 10(n - x + y)$$
$$= -11x + 10y + 11n + 50,$$
$$C_{XY} = 10x + (y + 10) + 10(n - x + y)$$
$$= 11y + 10n + 10.$$

Equilibrium will be reached when all three cost functions are equal, with no route being better than any other. Setting $C_X = C_Y$ and $C_X = C_{XY}$ results in the two equations $2x - y = n$ and $11x - 12y = 10n - 40$, which have solutions

$$x = \frac{2(n + 20)}{13} \quad \text{and} \quad y = \frac{80 - 9n}{13}.$$

It must be that $x \leqslant n$ and $0 \leqslant y \leqslant n$ for the solutions to make sense, and these inequalities reduce to $n \geqslant \frac{40}{11}$ and $\frac{40}{11} \leqslant n \leqslant \frac{80}{9}$, respectively.

So, provided that $\frac{40}{11} \leqslant n \leqslant \frac{80}{9}$ we have meaningful solutions for the equilibrium position. Finally, substituting these values of x and y back into the original cost functions results in the equilibrium position of

$$C_X = C_Y = C_{XY} = \frac{31n + 1010}{13}.$$

Recall that, with no relief road, the cost equilibrium function $C = \frac{1}{2}(11n + 100)$. Braess's Paradox emerges when $C_X > C$, which is when $\frac{1}{13}(31n + 1010) > \frac{1}{2}(11n + 100)$, and this means that again we need $n < \frac{80}{9} = 8\frac{8}{9}$.

Braess's Paradox will appear for any n which allows the equilibrium to be realized: $n = 4, 5, 6, 7, 8$.

In fact, Braess took $n = 6$ to demonstrate the point, which caused the traffic flow to be as in figure 4.5 and the equilibrium cost of travel before the relief road was built to be

$$C = \frac{1}{2}(11 \times 6 + 100) = 83$$

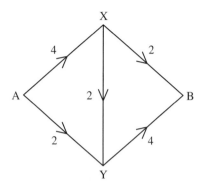

Figure 4.5.

and the cost afterwards to be

$$C_X = C_Y = C_{XY} = \tfrac{1}{13}(31 \times 6 + 1010) = 92.$$

The building of the extra road has made things worse!

We can also see that the invisible hand has indeed lost its grip. Now the average cost is the more formidable expression

$$A = \frac{1}{n}(x \times 10x + (n - x)(n - x + 50) + y(y + 10)$$
$$+ (x - y)((x - y) + 50) + (n - x + y) \times 10(n - x + y)),$$

which simplifies to

$$A = \frac{1}{n}(12y^2 + (20n - 22x - 40)y + 22x^2 - 22nx + 11n^2 + 50n)$$

and which we will consider as a function of y for any given x, again completing the square to get

$$A = 12\left[\left(y + \frac{10n - 11x - 20}{12}\right)^2 - \left(\frac{10n - 11x - 20}{12}\right)^2\right]$$
$$+ 22x^2 - 22nx + 11n^2 + 50n$$
$$= 12\left(y + \frac{10n - 11x - 20}{12}\right)^2 - 12\left(\frac{10n - 11x - 20}{12}\right)^2$$
$$+ 22x^2 - 22nx + 11n^2 + 50n.$$

Now we have isolated y we can see that A will achieve its minimum value when $y = \tfrac{1}{12}(11x - 10n + 20)$ for any given x.

In the selfish case, we saw that equilibrium is achieved when $x = \frac{2}{13}(n + 20)$, which would now make $y = \frac{1}{39}(175 - 27n)$, which we can compare with the original $y = \frac{1}{13}(80 - 9n)$. The invisible hand has indeed disappeared.

There now exist any number of formulations of the phenomenon, using networks which are ever more complicated, with cost functions linear or otherwise. As a first step, in the original Braess form that we have looked at we might imagine that making XY two-way would make a difference but, although the numbers change, the paradox stubbornly reemerges.

We have framed the paradox, as Braess originally framed it, in terms of traffic flow and we have mentioned two cases where it might have a real impact on a road system. The article 'The prevalence of Braess's Paradox' by R. Steinberg and W. I. Zangwill in the journal *Transportation Science* (17:301–18, 1983) provides more examples. That said, the paradox is essentially about flow in a network in which there are varied cost functions associated with its arcs, and as such it is not restricted to traffic flow: examples of it appear in cases of water flow, computer data transfer, mechanical and electrical networks and telephone exchanges. In 1990 the British Telecom network suffered in such a way when its 'intelligent' exchanges reacted to blocked routes by rerouting calls along 'better' paths. This in turn caused later calls to be rerouted with the cascade effect leading to a catastrophic change in the network's behaviour.

Braess's Paradox is a manifestation of small local changes unpredictably resulting in large global effects and has become a force to be reckoned with.

Chapter 5

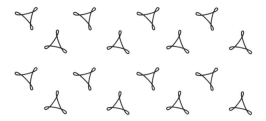

THE POWER OF COMPLEX NUMBERS

Imagine a person with a gift of ridicule. [He might say] First that a negative quantity has no logarithm; secondly that a negative quantity has no square root; thirdly that the first nonexistent is to the second as the circumference of a circle is to the diameter.

Augustus De Morgan

Strange Happenings

Apart from admiration for the committed effort and ingenuity displayed in a note to the *American Mathematical Monthly* by H. S. Uhler, the casual reader might be surprised by the approximate value

0.207 879 576 350 761 908 546 955 619 834 978 770 033 877 841 631 769 614

the author gives to $i^i = \sqrt{-1}^{\sqrt{-1}}$: a real number.[1] The purpose of the note was to give high-order decimal approximations to eight numbers, each of which is a power of e; two of those numbers

---

[1]H. S. Uhler, 1921, On the numerical value of i^i, *American Mathematical Monthly* 28(3):114–16.

were e^π, the Gelfond constant, which we will mention later, and $e^{-\pi/2}$, which was written alternatively as i^i. The remarkable fact that $\sqrt{-1}^{\sqrt{-1}} = e^{-\pi/2}$ together with a look at De Morgan's quotation above and some material underlying them will occupy us over the next few pages.

Throughout their history, complex numbers have caused conceptual and philosophical difficulty, and their logarithms grave confusion. For example, in the mathematical ferment of the eighteenth century it was of natural and considerable importance to reconcile the two different answers to the integral (omitting the arbitrary constant)

$$\tan^{-1} x = \int \frac{1}{x^2 + 1}\, dx = \int \frac{1}{(x + i)(x - i)}\, dx$$
$$= \frac{1}{2i}\int \frac{1}{x - i} - \frac{1}{x + i}\, dx = \frac{1}{2i}\ln \frac{x - i}{x + i},$$

which meant giving a sensible meaning to the logarithm of complex numbers, and, in particular, to $\ln i$. Johann Bernoulli argued that, since $(-x)^2 = x^2$, $\ln(-x)^2 = \ln x^2$ and so $2\ln(-x) = 2\ln x$, which means that $\ln(-x) = \ln x$, and in particular $\ln(-1) = \ln 1 = 0$; this of course would mean that $\frac{1}{2}\ln(-1) = \ln \sqrt{-1} = \ln i = 0$. The great Leibniz disagreed, arguing that, if $y = \ln x$, then

$$x = e^y = 1 + y + \frac{y^2}{2!} + \frac{y^3}{3!} + \cdots$$

and putting $x = -1$ and $y = 0$ would result in $-1 = 1$. It was Leibniz's belief that $\ln(-1)$ had to be an imaginary number and it was the genius Leonard Euler's arguments that were to find in his favour and also resolve the seeming contradiction. In a 1728 letter to Bernoulli he provided arguments that easily led to the expression $\frac{1}{2}\pi = -\sqrt{-1}\ln \sqrt{-1}$ and in his 1749 article (published in 1751), 'Recherches sur les racines imaginaries des equations' (*Memoires de l'Academy de Science de Berlin* 5(1749):222–88), in attempting to prove that the complex numbers were *complete*, and as a special case of a general formula, he evaluated $\sqrt{-1}^{\sqrt{-1}}$ as 0.207 879 576 350 7, remarking

Qui est d'autant plus remarquable, qu'elle est réelle, et qu'elle renferme même une infinite de valeurs réelles différentes.

(What is all the more remarkable is that it is real, and it has an infinity of different actual values.)

It was the multivalued nature of the logarithm of complex numbers that was the key to resolving the Bernoulli–Leibniz conundrum.

Chapter 7 of *Nonplussed!* referred to the distinguished eighteenth-century British mathematician Augustus De Morgan (actually born in India) and the *paradoxers* who plagued him. This chapter's opening quotation might suggest that paradoxers had been at work, but De Morgan knew better and offered his observation as an amusing example of the subtlety which is inherent in the study of complex numbers.

Complex numbers are, perhaps, well named, although *subtle numbers* might be a better alternative: at least the phrase would sidestep the suggestion of frightening difficulty and hint at the inherent beauty of this 'ultimate' extension of the number system. (The use of the quoted adjective must be tentative, with the existence of Hamilton's *quaternions*, Conway's *surreal numbers* and Cantor's *transfinite numbers*, etc.

The 'Ultimate' Extension

The fundamental need to solve equations has brought with it successive, natural extensions of the natural number system $\mathbb{N} = \{1, 2, 3, \dots\}$.

For a simple equation like $x + 2 = 5$, where all of the coefficients belong to $\mathbb{N}$, there is no difficulty; its solution is $x = 3$, and $3 \in \mathbb{N}$: an equation which exists in $\mathbb{N}$ has its solution in $\mathbb{N}$.

Now consider the following alternatives, each of which is an equation framed in $\mathbb{N}$, but in order to find solutions(s) we need to move to an extended number system:

- $x + 5 = 2$: to solve this we need to extend $\mathbb{N}$ to the set of all integers $\mathbb{Z} = \{\dots, -3, -2, -1, 0, 1, 2, 3, \dots\}$, or *Zahlen*, the German word for *number*;

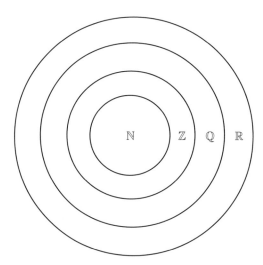

Figure 5.1.

- $5x = 2$ requires a further extension to $\mathbb{Q} = \{\text{Fractions}\}$, or *quotients*, for its solution;
- $x^2 = 2$ requires the extension to the real numbers $\mathbb{R}$, and so include the irrationals like $\pm\sqrt{2}$, and we have captured its two solutions.

These extensions are represented by the schema shown in figure 5.1. Each progressively fills up the number line, which we can picture as the x-axis, and once $\mathbb{R}$ has been rigorously defined (and that is not at all a simple matter), it can be shown that the number line is full: there is no more room for more extensions to solve more equations.

So, what about the equation $x^2 = -1$?

The long, tortuous and mathematically painful story of the extension to the complex number system $\mathbb{C}$ cannot occupy us here, but the inclusion of a single new number $i = \sqrt{-1}$ brings about the more general numbers $x + iy$, for $x, y \in \mathbb{R}$, and extends figure 5.1 by one final, enclosing circle. It proves to be the final extension that is necessary to solve all polynomial equations. To be precise, *any polynomial of degree n with coefficients in $\mathbb{C}$ has all of its n solutions in $\mathbb{C}$*; a result of such importance that it is universally known as the *Fundamental Theorem of Algebra*, first proved in 1799 by the inimitable Gauss.

Early Problems

As the number system is progressively extended—natural numbers, integers, rationals and reals—nothing very surprising happens. This is not to say that the rigorous definitions of some of these extensions are not themselves challenging, but more that the new numbers behave themselves 'properly'. The 'final' extension to the complex numbers, $\mathbb{C}$, brings with it beauty, great utility and the genesis of counterintuitivity. We will mention two early problems.

First, the concept of order is lost; that is, it is no longer possible to ask the seemingly reasonable question, 'which of two numbers is the greater?' To see this, since $i \neq 0$, we can ask the question, is $i > 0$ or is $i < 0$? If $i > 0$ we can multiply both sides of the inequality by i and preserve it to give $i^2 > 0i$, this means that $-1 > 0$. Alternatively, it must be that $i < 0$ and this time multiplying both sides by i it must therefore reverse the inequality to give $i^2 > 0i$ once more and again $-1 > 0$. The assumption of order brings with it an irreconcilable contradiction and so must be abandoned.

To approach the second difficulty, we need to consider the *Fundamental Theorem of Arithmetic*: any natural number can be factorized in a unique way into the product of primes. For example, 504 can be decomposed into the product $2^3 \times 3^2 \times 7$ and no other combination of primes can be chosen. Notice that it is a theorem—and so needs proof. To be exact, it is two theorems:

(1) Any natural number can be factorized into primes.
(2) The factorization is unique (which denies 1 being considered as a prime).

The equivalent result appears in Book VII of Euclid's *Elements* as a combination of three propositions:

Proposition 30. If two numbers multiplied by one another make some number, and any prime number measures the product, then it also measures one of the original numbers.

Proposition 31. Any composite number is measured by some prime number.

Proposition 32.   Any number is either prime or is measured by some prime number.

Whatever the provenance of the result, it is important and often taken for granted—but its veracity is entirely dependent on the nature of the natural numbers and its associated set of prime numbers. We can easily change matters. Take as an example the set S of natural numbers of the form $3n + 1$, $n = 0, 1, 2, \ldots$; that is, $S = \{1, 4, 7, 10, \ldots\}$. This set is *closed* under multiplication; that is, the product of two such numbers is itself such a number. It also has its set of prime numbers: for example, 4, 10 and 25 have no factors within the system and so must be considered as prime; yet $100 = 4 \times 25 = 10 \times 10$.

Now consider the set of complex numbers $a + b\sqrt{-5}$, where a and b are real numbers, and the associated set of (what are called *algebraic*) integers $a + b\sqrt{-5}$ for a and b ordinary integers. It is the case (and it is not at all obvious) that 2, 3, $(1 + \sqrt{-5})$ and $(1 - \sqrt{-5})$ are all primes in the system, yet

$$6 = 2 \times 3 = (1 + \sqrt{-5})(1 - \sqrt{-5}).$$

Unique factorization by primes is no longer valid.

Now that we have complex numbers properly placed and our mind receptive to lurking difficulty, we will consider what should be a simple computation for a calculator.

A Calculator's Dilemma

Calculators are becoming ever more sophisticated and for very little money we can buy one which is capable of giving many answers in exact form, and of dealing with expressions which are complex in both senses of the word. Such sophisticates should have no difficulty in dealing with $(-1)^{2/3}$ or $(-1)^{3/2}$ but even they might not tell the whole story—and as we write there are plenty of calculators in circulation which would simply announce 'Math Error' as their answer.

Using the standard laws of indices, we can achieve the evaluations with ease:

$$(-1)^{2/3} = ((-1)^{1/3})^2 = (-1)^2 = 1$$

and

$$(-1)^{3/2} = ((-1)^3)^{1/2} = (-1)^{1/2} = i.$$

Of course, we could have evaluated the expressions by splitting them in the alternative way with the first becoming

$$(-1)^{2/3} = ((-1)^2)^{1/3} = (1)^{1/3} = 1.$$

The symbol $(-1)^{2/3}$ really does seem to have the value 1. But now look at the result of doing this to the second expression:

$$(-1)^{3/2} = ((-1)^{1/2})^3 = (i)^3 = -i.$$

In a way we can provide a sort of explanation of this second phenomenon: $i = \sqrt{-1}$ for sure, but -1 has two square roots, $\pm i$. Interpreting the symbols one way round seems to choose one of the roots and the other way round the other root. But why? And what, if anything, has the evaluation of $(-1)^{2/3}$ to do with complex numbers?

What Is a^b?

If we wish to evaluate, for example, 3^4, we can write it as $3 \times 3 \times 3 \times 3$ and evaluate the expression to 81 (even without a calculator!). The likes of 273^{14} would have us reach for electronic help but there would be no surprise that the help was effective and that we have been saved the drudgery of performing thirteen long multiplications. But what about, for example, $2^{\sqrt{2}}$? Since we cannot write the number 2 down $\sqrt{2}$ times we reach for our calculator, press the buttons and magically it seems to be able to as it comes up with the answer to the degree of accuracy of which it is capable: $2.665\,144\,142\,690\,225\,188\,650\,297\,249\,87\ldots.$

In fact, this particular number has a famous history and a name to convince one of its importance: it is the Gelfond–Schneider constant and its nature was the subject matter of the seventh of the great David Hilbert's famous 1900 list of unsolved problems. It took until 1934 before Aleksandr Gelfond and Theodor Schneider independently proved that

> If $\alpha \, (\neq 0, 1)$ and β are algebraic numbers and if β is not a real rational, then any value of α^{β} is transcendental.

The calculator has done its best but is doomed to give a mere approximation to the actual value of the expression.

Of course, $2^{\sqrt{2}}$ is simply an example of an awkward power; we could ask the value of any other difficult expression: uninterestingly, $3.827^{4.916}$, or, very much more interestingly, e^{π} or π^{e}. In fact, the nature of π^{e} remains an unsolved problem, yet the Gelfond–Schneider theorem establishes the transcendence of e^{π}, even though e itself is transcendental—for reasons we shall see.

In fact, all exponential expressions, whether simple or complicated, are evaluated in the same manner, using the identity $a^{b} = e^{b \ln a}$, where both the exponential and logarithmic functions are evaluated using numerical methods based on the standard Taylor Series. A somewhat heavy approach for the likes of 3^{4}, one might think, but it does allow the standard calculator to cope with the formidable $2^{\sqrt{2}}$: what it might not cope with is $(-1)^{2/3}$ or $(-1)^{3/2}$ and the reason for this failure is exposed by the rewrites $(-1)^{2/3} = e^{(2/3)\ln(-1)}$ and $(-1)^{3/2} = e^{(3/2)\ln(-1)}$. We are forced to confront the same dilemma as those eighteenth-century luminaries: ascribing a value to $\ln(-1)$. Recall De Morgan's quip: 'a negative quantity, (which) has no logarithm'.

The Calculator Problem Solved

The standard representation of a complex number in rectangular and polar forms has us write $z = x + iy = r \cos \theta + ir \sin \theta = r(\cos \theta + i \sin \theta) = re^{i\theta}$. If we are to give a meaning to $w = \ln z$, then it must be that $e^{w} = z$, and choosing the rectangular form for w and the polar form for z we have $e^{u+iv} = re^{i\theta}$ and so $e^{u}e^{iv} = re^{i\theta}$, which means that $e^{u} = r$ and $v = \theta$. All of this results in the observation that $u = \ln r = \ln |z|$ and $v = \theta =$

arg z and the fact that $\ln z = \ln|z| + i\arg z$. Evidently, $z = -1$ is a complex number with $|-1| = 1$ and $\arg(-1) = \pi$ and using the above we have that

$$\ln(-1) = \ln|-1| + i\arg(-1) = 0 + i\pi = i\pi.$$

This means that

$$(-1)^{2/3} = e^{(2/3)\ln(-1)} = e^{(2/3)(\ln|-1|+i\arg(-1))}$$
$$= e^{(2/3)(0+i\pi)} = e^{(2/3)i\pi}$$
$$= \cos(\tfrac{2}{3}\pi) + i\sin(\tfrac{2}{3}\pi)$$
$$= -\tfrac{1}{2} + i\tfrac{\sqrt{3}}{2},$$

which is great, but it is not the 1 it evidently should be.

One way of achieving reconciliation is to look at the equation $z = (-1)^{2/3}$ and its equivalent form $z^3 = (-1)^2 = 1$. In asking for the value of $(-1)^{2/3}$ we are asking for a cube root of 1, and Gauss's Fundamental Theorem tells us that there are precisely three of them. Factorizing the cubic results in the equation

$$z^3 - 1 = (z - 1)(z^2 + z + 1) = 0,$$

which generates our long-desired $z = 1$ as well as

$$z = \frac{-1 \pm \sqrt{1-4}}{2} = \frac{-1 \pm \sqrt{-3}}{2} = \frac{-1 \pm i\sqrt{3}}{2}.$$

Another (more useful) approach is to realize that the polar form of a complex number is many valued. For example, $1 = e^{2\pi i} = e^{4\pi i} = e^{6\pi i} = \cdots$; in each case the modulus is 1 but the arguments differ by multiples of 2π. In our example above, we are dealing with the complex number -1, which has the polar forms $-1 = e^{i\pi} = e^{3i\pi} = e^{5i\pi} = \cdots$ and, taking the general case, we have that

$$(-1)^{2/3} = e^{(2/3)\ln(-1)} = e^{(2/3)(\ln|-1|+i\arg(-1))} = e^{(2/3)(0+i(2k+1)\pi)}$$

for any integral $k = 0, 1, 2, \ldots$. If we start to enumerate the possibilities, we have that

$$(-1)^{2/3} = e^{(2/3)(0+i(2k+1)\pi)}$$

$$= \begin{cases} e^{(2/3)i\pi} = \cos(\frac{2}{3}\pi) + i\sin(\frac{2}{3}\pi) \\ \qquad\quad = -\frac{1}{2} + i\frac{\sqrt{3}}{2}, & k = 0, \\[1em] e^{(6/3)i\pi} = e^{2i\pi} \\ \qquad\quad = \cos(2\pi) + i\sin(2\pi) = 1, & k = 1, \\[1em] e^{(10/3)i\pi} = \cos(\frac{10}{3}\pi) + i\sin(\frac{10}{3}\pi) \\ \qquad\quad = -\frac{1}{2} - i\frac{\sqrt{3}}{2}, & k = 2, \end{cases}$$

with the natural value appearing in the second row and two alternatives either side of it. Of course, there are an infinite number of choices for $\arg(-1)$ but that Fundamental Theorem holds sway since moving to the next value of $k = 4$,

$$(-1)^{2/3} = e^{14i\pi/3} = e^{12i\pi/3} \times e^{2i\pi/3} = 1 \times e^{2i\pi/3},$$

starts to repeat the values.

A Remarkable Result

Now that the evaluation of powers has been discussed and an anomaly explained, we move to the meaning of i^i.

Since $|i| = 1$ and the fundamental value of $\arg(i) = \frac{1}{2}\pi$, we have that

$$i^i = e^{i\ln i} = e^{i(\ln 1 + i\pi/2)} = e^{-\pi/2},$$

and we have that striking result: $i^i = \sqrt{-1}^{\sqrt{-1}} = e^{-\pi/2}$.

That said, we know that the full story has not been told, since $\arg(i) = (\frac{1}{2}\pi + 2k\pi)$.

The complete form of the expression is

$$i^i = e^{i\ln i} = e^{i(\ln 1 + i(\pi/2 + 2k\pi))} = e^{-(\pi/2 + 2k\pi)} \quad \text{for all } k = 0, 1, 2, \ldots.$$

So, the exotic $\sqrt{-1}^{\sqrt{-1}}$ has an infinite number of values, each of which is real and each of which involves the ubiquitous e and π.

Now, perhaps, the reader can cope with the puzzling

$$-1 = -1, \qquad \sqrt{-1} = \sqrt{-1}, \qquad \sqrt{\frac{-1}{1}} = \sqrt{\frac{1}{-1}},$$

$$\frac{\sqrt{-1}}{\sqrt{1}} = \frac{\sqrt{1}}{\sqrt{-1}}, \qquad (\sqrt{-1})^2 = (\sqrt{1})^2, \qquad -1 = 1.$$

And we should conclude by addressing the De Morgan quotation with which the chapter began. His words translate into symbols as

$$\frac{\ln(-1)}{\sqrt{-1}} = \frac{\ln|-1| + i\arg(-1)}{i} = \frac{i\pi}{i} = \pi = \frac{2\pi r}{2r} = \frac{C}{D},$$

at least as a principal value!

Chapter 6

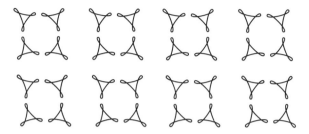

BUCKING THE ODDS

It is better to do the right problem the wrong way than to do the wrong problem the right way.

Richard Hamming

In this chapter we will consider two problems, each of which caused large-scale consternation and disbelief when they came to the attention of the public. The first had academic origins, the second was inspired by a popular American television show.

The Three-Hat Problem

We have already considered matters relating to red- and blue-hat wearers not knowing the colour of the hat each is wearing. Chapter 1 had a group of them sitting listening to the chiming of a clock, waiting for revelation. Here we will give each of them a more active role: guessing the colour of their hat, but under the following conditions.

- The players act as a team. The *team* wins or loses, not individuals.

- When the hats have been placed on the heads there must be no communication between team members.
- All must answer simultaneously.
- Each is allowed to pass rather than guess a colour.
- The team wins if at least one player guesses correctly and none guess incorrectly. Otherwise, or if they all pass, it loses.

Note that it is perfectly permissible for the team to discuss a strategy before the placement of the hats; the question is, what strategy will maximize their chance of success?

Strategies

Initially, we will look at the problem in its original form, where there are three players in the team: A, B and C.

Strategy 1. Each chooses red or blue randomly. With eight possible triplets of colours this would result in the probability of success of $\frac{1}{8} = 0.125$, with each team member having to guess correctly.

Strategy 2. Include the possibility of a pass and randomly choose one of the three possibilities red, blue or pass. Here the calculation is a little more delicate and it is best to separate the $3^3 = 27$ possibilities according to whether the combination has 3, 2, 1 or 0 passes. The eight possible triplets of hats that we have mentioned above are, of course,

RRR,   RRB,   RBR,   RBB,   BRR,   BRB,   BBR,   BBB.

If all players pass, PPP, the team loses.

If there are two passes, they can occur in six ways, with the other hat coloured either red or blue. In every case there are four possibilities when the other player guesses correctly. For example, PPR wins if the hat combination happened to be RRR, RBR, BRR or BBR.

If there is one pass, it can occur in twelve ways among the remaining red and/or blue hats. In every case there are two possibilities when the other players both guess correctly. For example, PRR wins if the hat combination happened to be RRR or BRR.

If there are no passes, we revert to strategy 1.

Putting all of this together, the probability of the team winning is

$$\tfrac{1}{27} \times 0 + \tfrac{1}{27} \times 6 \times \tfrac{4}{8} + \tfrac{1}{27} \times 12 \times \tfrac{2}{8} + \tfrac{1}{27} \times 8 \times \tfrac{1}{8} = \tfrac{7}{27} = 0.259.$$

Strategy 3. Agree that two of the team pass and the third guesses randomly. The probability of team success is clearly $\tfrac{1}{2} = 0.5$.

Things improve as we change strategies and we have reached the level of a single random guess, but here is where matters take a surprising turn. It does not seem possible that looking at the others' hat colours can yield any useful information and, therefore, there can be no prior strategy that can increase the possibility of success above that of random chance: but then, Hamming Codes do not seem to have much to do with hat colours.

Genesis and Malachi

The problem is attributed to Dr Todd Ebert, a computer science instructor at the University of California at Irvine, who included it in his PhD thesis while studying at the University of California at Santa Barbara in 1998. He resurrected it as a problem offering extra credit to his students for solving a seven-player version, which he called the 'seven prisoners' puzzle'. From there it passed to the internet and to the science section of the *New York Times* (on 10 April 2001) and so to the world as a whole. Interest was shown by many university mathematics departments and major corporations, one of which, Bell Labs of Lucent Technologies, had Dr Peter Winkler as director of fundamental mathematics research. Dr Winkler met Dr Elwyn Berlekamp, then Professor of Mathematics and of Electrical Engineering and Computer

Table 6.1.

A	B	C	A	B	C	Outcome
Red	Red	Red	GBW	GBW	GBW	Lose
Red	Red	Blue	Pass	Pass	GBC	Win
Red	Blue	Red	Pass	GBC	Pass	Win
Red	Blue	Blue	GRC	Pass	Pass	Win
Blue	Red	Red	GBC	Pass	Pass	Win
Blue	Red	Blue	Pass	GRC	Pass	Win
Blue	Blue	Red	Pass	Pass	GRC	Win
Blue	Blue	Blue	GRW	GRW	GRW	Lose

GBW, guess blue: wrong; GBC, guess blue: correct; GRW, guess red: wrong; GRC, guess red: correct.

Science at the University of California at Berkeley, at a conference in New Orleans. It was Berlekamp's extensive expertise in coding theory that caused him to make the connection that was to lead to the remarkable resolution that it is possible for the three-person team to achieve a probability of success of $\frac{3}{4}$ and the general n-person team that of $n/(n + 1)$.

So what is this strategy for three players that Elwyn Berlekamp took about half an hour to find? It was, for each member of the team, as follows:

1. If you see two hats that have the same colour, guess the other colour.

2. If you see two hats of different colour, pass.

Table 6.1 details the strategy for each player and gives the outcome in each of the possible eight cases. In spite of the individuals guessing correctly six times and incorrectly six times, the group accumulates six wins and two losses, giving the probability of winning as $\frac{6}{8} = \frac{3}{4} = 0.75$.

Notice that the correct guesses are sparsely spread out (recall that we only need one of them for team success) and that the incorrect guesses are bunched together. Put succinctly, to increase the chances of success the team must adopt the somewhat counterintuitive strategy of being *wrong together, not correct together*.

This strategy for three players is easy enough to implement, but how did Elwyn Berlekamp come up with it so quickly and why did Todd Ebert test his students with a seven-player variant?

The Seven-Hat Problem

With three hats the strategy is meaningful since among any triplet of hats there must be at least two the same colour and this event is the trigger for action. With seven hats we are assured that at least four of them will have the same colour and we could use the equivalent trigger to adapt the strategy to:

1. If you see four hats that have the same colour, guess the other colour.
2. Otherwise, pass.

To see how this strategy affects the team's chances of success, suppose that four hats are coloured red, then all of the red-hatters will pass and each of the other three players will choose blue. Therefore, the team will win only if there are precisely four red hats and three blue hats.

The four red hats can be distributed among the seven places in $\binom{7}{4}$ ways and, taking into account that they could have been blue, we have that the probability of the team winning is

$$\frac{2 \times \binom{7}{4}}{2^7} = \frac{35}{64} = 0.55.$$

This is better than the single, random guess strategy but it certainly is not the

$$\frac{7}{7+1} = \frac{7}{8} = 0.875$$

which was heralded earlier. This is where Hamming Codes come in.

Protective Codes

Codes are easily associated with the desire to transmit messages between friends which cannot be read by an intercepting enemy;

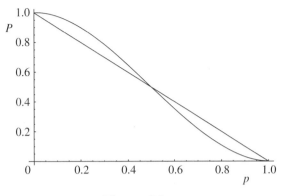

Figure 6.1.

one immediately thinks of war, government agencies and commercial companies. Yet, there are other codes too, which are just as crucial in their own way: error-protecting codes. Electronic transmission of binary data is not error free and the desire to have our data protected is a reasonable one and as a consequence, many codes have been invented to provide that protection. They take two forms: error detecting or error correcting; the former will flag a transmission error, the latter automatically correct it. Here we are interested in the latter type.

One key to the various methods of error correction is to build in redundancy and to use the technique of *nearest-neighbour decoding*, which takes a received, corrupted vector to the codeword nearest to it. An illustrative example is provided by the *binary repetition codes*. If we wanted to transmit the single digits 0 and 1, we could do so simply by sending 0 as 0 and 1 as 1, which would not cope at all with corruption. Alternatively, we could consider all triplets of binary digits

(000 001 010 100 : 111 110 101 011)

two of which, (000 111), are our codewords for 0 and 1 respectively; that is, if we wish to transmit a 0, we actually transmit 000 and to transmit 1 we send 111. If a single error occurs during transmission, we know that the received vector is incorrect and we replace it by the nearer of the two codewords, to the left of the colon for 0 and to the right for 1. So, we are sending more data but enjoy the benefit of the error protection. To judge the level of improvement, if p is the constant (small) probability of corruption of a single digit (i.e. transforming 0 to 1 or vice versa), then we will decode a message correctly if it incurs no errors or just one error. The probability of this is

$$P = (1 - p)^3 + 3p(1 - p)^2 = (1 - p)^2(1 + 2p),$$

which should be compared with $P = 1 - p$ for the single-digit transmission. Figure 6.1 shows the plot of the two functions and shows that we have an improvement (since p is assumed small). For the method to work it is crucial that the set of eight possible received vectors is partitioned into the two subsets of four; the one representing the codeword and its three satellite nearest neighbours. We can see that this will happen for any odd number of repetitions but no even number (what is the nearest neighbour to 01 or to 0011?) and binary repetition does provide the protection we seek—but it is not very efficient.

At the end of World War II Richard Hamming moved from the Manhattan Project to Bell Telephone Laboratories, where he worked with both Claude Shannon and John Tukey, from which collaboration the vast and crucial subject of information theory came into being. Hamming was concerned with data integrity in the ancient IBM computers that he was bound to use and developed what have become known as Hamming Codes with a view to protecting his programs and data (it remains a good choice for randomly occurring errors in modern computer RAM).

An $H(n, d)$ code uses binary vectors of length n to code data of length d. This means that there are 2^d codewords protected by $2^n - 2^d$ vectors. Our binary repetition code is really a $H(3, 1)$ code, but here our interest lies with the $H(7, 4)$ code, a version of which has codewords

$$\begin{pmatrix} 0000000 & 0001011 & 0010111 & 0011100 \\ 0100101 & 0101110 & 0101110 & 0111001 \\ 1000110 & 1001101 & 1010001 & 1011010 \\ 1011010 & 1101000 & 1110100 & 1111111 \end{pmatrix}.$$

All vectors differ from each other in three places; it can therefore detect up to three errors and it can also correct a single error, which is all we need. The $2^7 = 128$ possible vectors are divided into sixteen lots of eight: the appropriate codeword and the seven satellite vectors which differ from it in precisely one place. For example, the top left codeword of 0000000 yields

$$\begin{pmatrix} 0000000 & 0000001 & 0000010 & 0000100 \\ 0001000 & 0010000 & 0100000 & 1000000 \end{pmatrix}.$$

Hamming and Hats

If we have n hat wearers and code a red hat by a 0 and a blue hat by a 1, any hat configuration can be thought of as a vector of length n. If we consider the Hamming Code of that length, the hat configuration may be represented by a codeword or a satellite, and that fact determines the strategy to use.

If we consider the original three-hat puzzle and the $H(3,1)$ code above, we can make the following transition.

The original strategy

1. If you see two hats that have the same colour, guess the other colour.

2. If you see two hats of different colour, pass.

becomes

1. If you see a vector which could be a codeword, choose your digit to ensure that it is not a codeword.

2. Otherwise, pass.

If it happens that the chosen vector is a codeword, say 000 (i.e. all hats are red), then each of the three players will see the possibility and choose so as to avoid the codeword and say blue (i.e. a 1): all three will be wrong in one place. If it happens not to be a codeword, only one player will think that it could be; he will choose the appropriate nearest neighbour and be correct and the other two will pass: the team wins.

For their extra credit, Todd Ebert's students would have had to consider the $H(7,4)$ code and exactly the same translation.

Strategy.

1. If you see a vector which could be a codeword, choose your digit to ensure that it is not a codeword.
2. Otherwise, pass.

Exactly the same reasoning as before ensures that all choose incorrectly if the configuration is a codeword and otherwise all pass but one, who avoids the codeword and chooses the appropriate nearest neighbour to it. This means that the probability of the team winning is

$$\frac{2^7 - 2^4}{2^7} = \frac{7}{8}.$$

And finally, why were there seven hatters? For this approach to work we need the partitioning of the vectors as we have had them: for a given number of hat wearers (length of codeword) n, the set of 2^n vectors formed from 0s and 1s needs to divide into clusters, each of which contains a codeword at its centre and the n vectors which differ from it in a single place. The clusters, therefore, have $n + 1$ vectors in each of them and therefore $n + 1$ must be a factor of 2^n and so itself be a power of 2, let us say 2^m. Therefore, $n + 1 = 2^m$ and $n = 2^m - 1$. Notice that, if the number of codewords is C, then we have that $(n + 1)C = 2^n$ and so $2^m C = 2^n$ and $C = 2^{n-m}$.

With $m = 2$ we have the original three-hat problem and with $m = 3$ we have the problem for extra credit. (Actually, it is possible to adapt the method for other values of n.) In general, with n hat wearers we must consider $H(n, n - m)$, which has 2^{n-m} codewords and 2^n vectors, so the probability of the team being successful is

$$\frac{2^n - 2^{n-m}}{2^n} = 1 - \frac{1}{2^m} = \frac{2^m - 1}{2^m} = \frac{n}{n + 1}.$$

To solve our intriguing problem we have touched the surface of a vast and vastly important area of mathematical application, deliberately avoiding the associated definitions and results. We needed Hamming Codes and for many, often hidden purposes, we need others too. Here are just three examples. A *Reed–Muller* code was used to send photographs back from Mars by Mariner 9 in 1972; it has 64 codewords, each of length 32, and all are a distance 16 from each other; it can correct up to seven errors. *Cross-leaved Reed-Solomon* codes are used to protect information on CDs. They can correct 4000 consecutive errors, corresponding to a 2.5 mm long scratch. And the lowly, easily ignored, length 10 *ISBN* code can detect a single error or a single transposition in a book's code number.

Now we move to the second example in which a probability is not what it 'should' be.

Let's Make a Deal

We referred to the *Ask Marilyn* column of *Parade Magazine* in chapter 2. The 9 September 1990 issue had Marilyn respond to a reader's query inspired by the television game show *Let's Make a Deal*, which for many years had been hosted by Monty Hall: her answer to the query brought her nearly 10 000 responses from readers, most of them disagreeing with her; a number were from mathematicians and scientists whose responses ranged from hostility to disappointment at the nation's lack of mathematical skills. In 1991 the *New York Times* published a large front page article in a Sunday issue which declared:

> Her answer...has been debated in the halls of the C.I.A. and the barracks of fighter pilots in the Persian Gulf. It has been analyzed by mathematicians at M.I.T. and computer programmers at Los Alamos National Laboratory in New Mexico. It has been tested in classes ranging from second grade to graduate level at more than 1,000 schools across the country.

More recently, the CBS drama series *NUMB3RS* featured it in its 13 May episode of 2005, entitled 'Manhunt', and the *Financial Times* published a column about it on 16 August 2005, written by John Kay; this resulted in the publication of several letters on its 'Leaders and Letters' page on 18 and 22 August, and two follow-up columns on 23 August ('So you think you know the odds') and 31 August ('The Monty Hall problem—a summing up') in which it was acknowledged that there had been 'a large correspondence'. To put matters into a stark perspective, Kay noted that the late great Paul Erdős (reputedly) died still musing on it!

The whole confusing matter has understandably become associated with Monty Hall and is indeed commonly known as *The Monty Hall Problem*, which states:

> Suppose that you are on a game show, and that you are given the choice of three doors. You know that behind one door is a car and behind the others, goats. You are allowed to pick any single door and keep whatever lies behind it. Now suppose that you pick door number 1, but do not yet open it. The host, who knows what is behind all of the doors, then opens door number 2, which he knows has the remaining goat behind it. He then says to you, 'Do you want to swap to door number 3 or keep your first choice of door?'

The question is: Does it matter if you stick or you switch? Marilyn's answer was that the contestant should switch doors.

Surely, since the contestant knows that behind at least one of the doors there is a goat, the revelation of the goat provides no further information: it will make no difference whether one switches or not. So many, many people thought.

In fact, the problem was not a new one.

As with so many intriguing puzzles Martin Gardner had discussed a version of it (the three prisoners' problem) in his *Scientific American* column (in 1959) and again in *The Second Scientific American Book of Mathematical Puzzles and Diversions*, which first appeared in 1961. In his *Aha! Gotcha* book (published in 1982) he describes the following *Three Shell Game* variant:

> *Operator:* Step right up, folks. See if you can guess which shell the pea is under. Double your money if you win.

> After playing the game a while, Mr Mark decided he couldn't win more than once out of three.

> *Operator:* Don't leave, Mac. I'll give you a break. Pick any shell. I'll turn over an empty one. Then the pea has to be under one of the other two, so your chances of winning go way up.

Table 6.2.

Original door selection	A	A	A	B	B	B	C	C	C
Car location	A	B	C	A	B	C	A	B	C
Door Monty can open	B, C	C	B	C	A, C	A	B	A	A, B
Contestant does not switch	W	L	L	L	W	L	L	L	W
Contestant does switch	L	W	W	W	L	W	W	W	L

> Poor Mr Mark went broke fast. He did not realize that turning
> an empty shell had no effect on his chances. Do you see why?

The situation is not quite the same as Monty Hall in that the
player is not given the opportunity to change his selection; as
Gardner argues, the new information provided is valueless. Now
we will see what happens if that opportunity to switch is provided and therefore analyse what Gardner called 'a wonderfully
confusing little problem'.

Behind Closed Doors

Many analyses have been made of a problem which intrinsically
relies on conditional probability, but let us first simply list the
possibilities as shown in table 6.2, where we have labelled the
doors A, B and C. The penultimate row shows us that, if the
contestant does not switch, his probability of winning is $\frac{3}{9} = \frac{1}{3}$,
whereas the probability of his winning by switching is shown in
the last row to be $\frac{6}{9} = \frac{2}{3}$.

It is critical that Monty knows where the car is. If he does not,
he could open the door with the car behind it and, if under this
circumstance we consider the game void, modifying the above
table it is easy to see that it no longer matters whether the
contestant switches or not.

The formalists would, perhaps, be better satisfied by recourse
to the result of the eighteenth-century Presbyterian minister,
Thomas Bayes, whose solution to the problem of *inverse probability* appeared in his 'Essay towards solving a problem in
the doctrine of chances' of 1763, which was published posthumously in the *Philosophical Transactions of the Royal Society of
London*. It was in this essay that Bayes's theorem made its first

public appearance and *reversing the conditional* became part of mathematical literature. At the time it was well understood that, given an urn containing a known number w of white and b black balls, the probability of drawing a white ball is $w/(w + b)$. The reverse problem which asks what can be said of an unknown distribution of white and black balls given that one or more balls had been drawn and their colour identified was quite another matter.

The conditional probability $P(X \mid Y)$ (the probability of X given Y) of two events X and Y is defined by the formula $P(X \cap Y) = P(X \mid Y)P(Y)$ (the probability of X and Y equals the probability of X given Y times the probability of Y) and the simplest form of Bayes's result (which is all we will need) is, given two events X and Y, then

$$P(Y \mid X) = \frac{P(X \mid Y)P(Y)}{P(X)}.$$

It is often useful to expand the bottom probability, in this case into three parts, defined in terms of three mutually exclusive events R, S and T:

$$P(X) = P(X \cap R) + P(X \cap S) + P(X \cap T)$$
$$= P(X \mid R)P(R) + P(X \mid S)P(S) + P(X \mid T)P(T).$$

In the case of Monty Hall we can define the following events:

A the event 'the car is behind door A',
B the event 'the car is behind door B',
C the event 'the car is behind door C',
M_A the event 'Monty opens door A', etc.

If door A is chosen initially by the competitor, then we know that Monty has a choice of doors B and C to open and we have

$$P(M_B \mid A) = \tfrac{1}{2}, \quad P(M_B \mid B) = 0, \quad P(M_B \mid C) = 1.$$

So,

$$P(M_B) = P(M_B \mid A)P(A) + P(M_B \mid B)P(B) + P(M_B \mid C)P(C)$$
$$= \tfrac{1}{2} \times \tfrac{1}{3} + 0 \times \tfrac{1}{3} + 1 \times \tfrac{1}{3} = \tfrac{1}{2}.$$

Now the contestant can stick or change. If he keeps to door A, his probability of winning the car is

$$P(A \mid M_B) = \frac{P(M_B \mid A)P(A)}{P(M_B)} = \frac{\tfrac{1}{2} \times \tfrac{1}{3}}{\tfrac{1}{2}} = \tfrac{1}{3},$$

whereas, if he switches to door C, the probability becomes

$$P(C \mid M_B) = \frac{P(M_B \mid C)P(C)}{P(M_B)} = \frac{1 \times \tfrac{1}{3}}{\tfrac{1}{2}} = \tfrac{2}{3}.$$

Variants on the theme are legion and we will look at a few natural extensions culled from the article 'Generalising Monty's dilemma', by John P. Georges and Timothy V. Craine, which appeared in the 1995 March/April issue of *Quantum Magazine* (5(4):16–21).

One Car and Many Goats

Consider the less appealing case for the contestant of there being n doors, behind one of which is the car and $n-1$ of which a goat.

If the contestant sticks with the original choice of door, the probability of winning the car is $1/n$. Now suppose that Monty opens a door, behind which is a goat. There are now $n-2$ doors which remain unopened and to calculate the probability of the contestant winning by switching doors we need to evaluate the expression

(Probability of a goat behind the first door)

$\times$ (Probability of a car behind the second door

given there was a goat behind the first door)

$$= \frac{n-1}{n} \times \frac{1}{n-2} = \frac{n-1}{n-2} \times \frac{1}{n} > \frac{1}{n},$$

since $(n-1)/(n-2) > 1$. It is always the case that the switching strategy is better than the sticking strategy and with $n = 3$ we resurrect the previous numbers.

Many Cars and Many Goats

Suppose now that there are n doors behind which there are c ($\geqslant 1$) cars and therefore $n-c$ goats. We will assume that the contestant has no idea of the value of c, in which case it is possible for Monty to show a car or a goat without revealing all.

Notice that, if Monty reveals a goat, then $1 \leqslant c \leqslant n - 2$, whereas, if he reveals a car, $2 \leqslant c \leqslant n - 1$. In either case the probability of winning a car by sticking is c/n.

Now suppose that Monty reveals a goat. If the contestant is to win using a switching strategy, it must be the case either that a goat was picked first and a car second or that a car was picked first and another car second.

The probabilities of these events are

$$\frac{n-c}{n} \times \frac{c}{n-2} \quad \text{and} \quad \frac{c}{n} \times \frac{c-1}{n-2},$$

which combine to a total probability of

$$\frac{n-c}{n} \times \frac{c}{n-2} + \frac{c}{n} \times \frac{c-1}{n-2} = \frac{c}{n(n-2)}(n-c+c-1)$$

$$= \frac{(n-1)\,c}{(n-2)\,n} > \frac{c}{n}.$$

Now suppose that Monty reveals a car. If the contestant is to win using a switching strategy, it must be the case either that a goat was picked first and a car second or a car was picked first and another car second.

The probability of each of these events is

$$\frac{n-c}{n} \times \frac{c-1}{n-2} \quad \text{and} \quad \frac{c}{n} \times \frac{c-2}{n-2}.$$

The probabilities of these events are

$$\frac{n-c}{n} \times \frac{c-1}{n-2} + \frac{c}{n} \times \frac{c-2}{n-2} = \frac{c-1}{n} \times \frac{n-c}{n-2} + \frac{c}{n} \times \frac{c-2}{n-2}$$

$$< \frac{c}{n} \times \frac{n-c}{n-2} + \frac{c}{n} \times \frac{c-2}{n-2} = \frac{c}{n}\left(\frac{n-c}{n-2} + \frac{c-2}{n-2}\right)$$

$$= \frac{c}{n}.$$

So, if Monty reveals a goat, the contestant should switch, but if he reveals a car the contestant should stay with the first choice.

Finally, we will consider a variant which has its own intrigue.

Multi-Stage Monty Hall Dilemma

The original game show had three doors from which the contestant could pick. The rules of the show gave the contestant two decision stages: the initial choice and then the decision of whether to stick with that choice or change. Now suppose there are four doors, with a car behind one of them, and a game structured in the following manner.

Monty Hall says:

> You select one of the doors, and I will open another door behind which is a goat. Then you decide whether you wish to stick with your original selection or switch to one of the remaining doors. I will then open another door behind which is a goat. Finally, once again you can decide whether or not to stick or switch to the only other remaining door.

Now the contestant had three decision stages: the first pick, the first stick–change decision and the second stick–change decision.

As an example, the following could take place.

First pick for the contestant. Contestant chooses door A so doors B, C and D are available to Monty. Monty opens door B and doors A, C and D are available to contestant.

Table 6.3.

Stage 1	Stage 2	Stage 3	Probability of winning the car
Pick	Stick	Stick	0.250
Pick	Switch	Stick	0.375
Pick	Stick	Switch	0.750
Pick	Switch	Switch	0.625

First stick–change decision for the contestant (who switches).
Contestant chooses door C so doors A and D are available to
Monty. Monty opens door D so doors A and C are available to
contestant.

Second stick–change decision for the contestant (who sticks).
Contestant chooses door C. The contestant has, in this case,
chosen to switch and then stick.

M. Bhaskara Rao of the Department of Statistics at the North
Dakota University has analysed this situation ('On a game-show
problem of Marilyn Vos Savant and its extensions', 1992, *American Statistician* 46:241–42) and, more generally, for *n* doors and
n − 1 decision points for the contestant. Table 6.3 summarizes
the results of his analysis for the four-door example above.

It would be easy, having accepted the optimal switch solution
in the basic Monty Hall dilemma, to assume that the contestant would do best by switching in both Stage 2 and Stage 3.
However, as table 6.3 shows, the counterintuitive solution to the
three-stage Monty Hall dilemma is to stick in Stage 2 and switch
in Stage 3. Generally, in a multi-stage Monty Hall dilemma, the
contestant should stick with the initial choice until the very last
stage and then switch.

The perplexing nature of the problem is touchingly recorded
in Mark Haddon's remarkable book, *The Curious Incident of the
Dog in the Night-time*, where he writes:

> It also shows that Mr Jeavons was wrong and numbers are
> sometimes very complicated and not very straightforward
> at all. And that is why I like the Monty Hall Problem.

And so our brief investigation into this most controversial of ideas is at an end—unless the reader is a bridge player. An observation that can be made is that, since two times out of three the contestant will choose a door with a goat behind it, two times out of three Monty will have no choice as to which door to open to reveal a goat. This observation translates to the bridge principle of *restricted choice*, which causes optimum play for the N–S holding of A J 10 7 6—5 4 3 2 to be *finesse both the J and the 10*, whereas, with the holding A Q 10 7 6—5 4 3 2, it is *finesse the queen and then play the ace.* We will leave it to the interested bridge-playing reader to see why and perhaps the late bridge expert Alan Truscott's *New York Times* bridge column of 4 August 1991 might be a starting point; in it he points out that he had explained the principle in bridge terms forty years earlier. But perhaps Zen philosophy provides the best answer: It makes no difference which you choose. If you desire to win, you have already lost.

Chapter 7

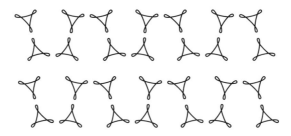

CANTOR'S PARADISE

Not everything that counts can be counted. Not everything that can be counted counts.

Albert Einstein

Jane Muir began the final chapter of her delightfully written book *Of Men and Numbers* with typically elegant prose:

There are times in history—the history of a man as well as a civilization—when one can look back and say, 'So this is where it's all been leading. It seems so obvious now, why didn't I realize before?' A man or a civilization comes to the end of a road; the journey is over; all the wanderings and travels down dead ends and over highways have led to this particular place and suddenly he realizes that he is at the end, the trip is over, the journey completed. Such was the feeling mathematicians had after Georg Cantor guided them over the last stretch of land.

They could rest. Their doubts and fears and wonders of where the road would lead were satisfied. But another road stretched out before them, an ill-defined, treacherous-looking path that both repelled and beguiled—and soon another journey began. The end of one trip suddenly became the beginning of another. Such, too, was the feeling mathematicians had after Georg Cantor opened their eyes to a new and foreign world.

Cantor's eminent contemporary Leopold Kronecker was more *repelled* than *beguiled* at this particular realization of his famous dictum *God made the integers, all else is the work of man.* We will look at the bare rudiments of Cantor's seminal ideas, ideas which followed paths from the obvious to the unbelievable, the elementary to the most profound.

To begin with, since we will make essential use of it, we will mention Euclid's fifth postulate and, since we are interested in that part of Cantor's work which contradicts the fifth common notion, we will mention that too.

Obvious Ideas

Euclid's *Elements* has claim to be the second bestselling book of all time (surpassed by the Bible). Written around 300 B.C. it chronicled much of the known mathematics of the time and is particularly (though by no means exclusively) regarded for its treatment of geometry, which begins with a set of twenty-three definitions, five postulates and five common notions; all seem at least very reasonable, if not self-evident. That said, the fifth postulate was at least cumbersome:

> That, if a straight line falling on two straight lines make the interior angles on the same side less than two right angles, the two straight lines, if produced indefinitely, meet on that side on which are the angles less than the two right angles.

It means that, in figure 7.1, if $a + b < 180$ the two lines will eventually meet.

In his *Commentary* on the *Elements*, the revered commentator Proclus (411–85) mentioned that the postulate was attacked from the outset and wrote 'this postulate ought even to be struck out of postulates altogether; for it is a theorem'. Among all of those definitions, postulates and common notions, it alone

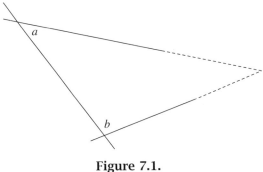

Figure 7.1.

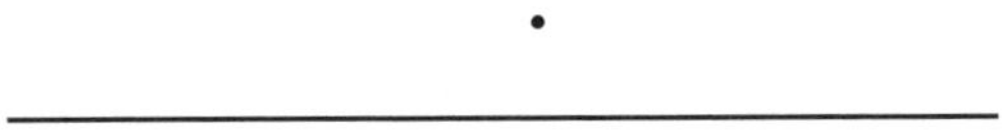

•

_____

Figure 7.2.

raised suspicion. The equivalent formulation, attributed to the eighteenth-century Scottish mathematician John Playfair (but known to Proclus) is even more disarming:

> Through a given point not on the line, there is one and only one line which can be drawn through that point parallel to the line.

Figure 7.2 shows the point and a finite segment of the line: the statement is surely obviously true.

It was not until Proposition 29 of Book 1 of the *Elements* that the fifth postulate was first used (and from there on used frequently in Book 1 and in later books):

> Proposition 29: A straight line falling on parallel straight lines makes the alternate angles equal to one another, the exterior angle equal to the interior and opposite angle, and the sum of the interior angles on the same side equal to two right angles.

Referring to figure 7.3, this means that $a = b, b = c$ and $b+d = 180$, respectively. The proof is brief and clear, but contains the statement:

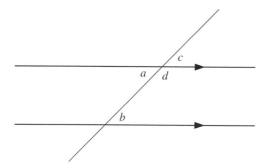

Figure 7.3.

But straight lines produced indefinitely from angles less than two right angles meet.

In fact, the great Proclus was wrong (as were many others): the fifth postulate is not a theorem but an independent statement, alternatives to which (there are no parallel lines or there is more than one parallel line through the point) are perfectly valid. It is a stark fact that those many propositions from 29 onwards (including the Pythagorean theorem) are true only of Euclidean geometry, defined as the geometry which arises from those precise definitions, postulates and common notions. Change the fifth postulate to 'no parallel lines' and we have spherical geometry, change it to 'more than one parallel line' and we have hyperbolic geometry, famously developed independently by the Russian mathematicians Nikolai Lobachevsky and János Bolyai. The model of spherical geometry is (unsurprisingly) the surface of a sphere (with 'straight lines' as the great circles); hyperbolic geometry fits less comfortably into our Euclidean space viewed through our Euclidean eyes: famous representations of it are the Klein–Beltrami disc and the Poincaré disc (explored with such wonder by the Dutch artist Maurits Escher) and the geometry of the pseudosphere. Here there are an infinite number of lines parallel to the given one, the Pythagorean theorem does not hold, the sum of the angles of a triangle is less than 180°, triangles with the same corresponding angles have the same area, not all triangles have the same angle sum and there are no similar triangles, etc. Jane Muir's words are as appropriate to this as they are to the work of Georg Cantor, to which we will now turn since it was he who called into question the fifth common notion, which, over the centuries, had itself seemed to be completely self-evident: 'The whole is greater than the part' or, moving to its commonly used Latin equivalent, 'Totem parte maius'.

It remained self-evidently true until the late nineteenth century, when Cantor's controversial work brought about results which confounded it, one of which so surprised him that, in an 1877 letter to his frequent correspondent, Richard Dedekind, he wrote, 'Je le vois, mais je ne le crois pas!' ('I see it, but I don't believe it!') We shall look at that result and some others of its type, but first we need a definition.

One-to-One Correspondence

Intuition, as is so often the case, will prove to be an unreliable guide and to counter its misleading ways we need a careful definition of how to compare the size of two infinite sets of objects, and that definition comes from one of the alternative ways we have of comparing two finite sets. If we have two bags of marbles and we are asked whether there is the same number of marbles in each bag, we could empty each bag and count the contents of each; if the two numbers tally, the bags did contain the same number of marbles. Yet, although this does answer the question, it does so by doing too much; we were not asked how many marbles were in each bag, just whether there is the same number. To answer the question directly we could repeatedly put one hand in each bag and remove marbles in pairs; if one bag empties before the other, they had different numbers of marbles inside them, otherwise the marbles were paired perfectly—or, using a more mathematical terminology, they were put in *one-to-one correspondence*. It is this idea of a one-to-one correspondence that Cantor used in dealing with the comparison of infinite sets. It is important to realize that the nature of the association matters not at all, it simply needs us to count off in pairs, just as the removal of marbles in each hand does so. For example, it is small surprise that the correspondence $n \to -n$ demonstrates the equivalence of the positive and negative integers but, having accepted that, the correspondence $n \rightarrow 2n$ brings about the equivalence of the positive integers and the even positive integers with equal alacrity; already Euclid's fifth common notion has been contradicted. In fact Cantor soon realized that it is characteristic of infinite sets that they do contradict the fifth notion and avoided powerful criticism by defining an infinite set as one which can be put into one-to-one correspondence with a proper subset of itself; any infinite set which can be put into one-to-one correspondence with the natural numbers was to become known as a 'countable' set and its 'size' (or *cardinality*) written as $\aleph_0$ (aleph null, or aleph zero).

The Rationals are Countable

With that definition surprises began to tumble from his pen. For example, using the property of unique factorization, if we consider the set

$$\mathbb{N}^n = \{(m_1, m_2, m_3, \ldots, m_n): m_1, m_2, m_3, \ldots, m_n \in \mathbb{N}\},$$

it can be put in one-to-one correspondence with an infinite subset of $\mathbb{N}$ by the association $(m_1, m_2, m_3, \ldots, m_n) \rightarrow 2^{m_1} \times 3^{m_2} \times 5^{m_3} \times \cdots \times (n\text{th prime})^{m_n}$. Dimension has nothing to do with size: $\mathbb{N}^n$ is countable; that is, the cardinality of $\mathbb{N}^n$ is $\aleph_0$.

If we take $n = 2$ in the above result and agree that the difference between (a, b) and a/b is merely notational, we can see that this establishes the fact that the rationals are countable. Yet another way of looking at a set's countability is to reason that its elements can be listed, but if we try to list them, then we naturally encounter the problem of missing out numbers; after all, between every two rationals there is another rational. How can we possibly list them in an exhaustive manner?

Cantor wondered this too and, in 1873, made a listing using a 'diagonal array'. Figure 7.4(a) displays the rationals as an infinite two-dimensional array with the first row consisting of those which have a numerator of 1, the second row those which have a numerator of 2, etc.

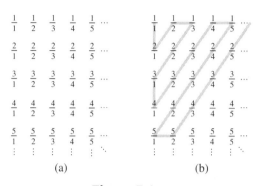

Figure 7.4.

Moving in the diagonal manner suggested in figure 7.4(b) brings about the following listing of the rationals:

$$\frac{1}{1}, \frac{2}{1}, \frac{1}{2}, \frac{1}{3}, \frac{2}{2}, \frac{3}{1}, \frac{4}{1}, \frac{3}{2}, \frac{2}{3}, \frac{1}{4}, \frac{1}{5}, \frac{2}{4}, \dots$$

Clearly, every rational number (repeatedly) appears (with, for example, $\frac{1}{1} = \frac{2}{2} = \frac{3}{3} = \cdots$) and we can extract from this list the distinct rationals simply by moving from left to right; in this way every rational number is counted once and only once and therefore put into one-to-one correspondence with the natural numbers.

The listing is made explicit, although it is a little ragged and it could be held to be unsatisfactory that the original list is populated with an infinite number of repeats which we have to sift through. Do 'natural' listings exist which dispense with this? The answer is 'yes' and two such are noted in the Neil Calkin and Herbert Wilf article 'Recounting the rationals' (2000, *American Mathematical Monthly* 107:360–64) as listed on *The On-Line Encyclopedia of Integer Sequences* website as sequences A038568 and A020651. A third is the subject of the article itself, which uses an elegant and beautifully elementary argument (highly related to Farey sequences and Stern–Brocot trees) to establish the fact. We will take the trouble to present it below.

The argument relies on a particular tree diagram, with the fraction $\frac{1}{1}$ as its top node and with the structure that each node a/b has two children:

- the left child, which is defined as $a/(a + b)$
- the right child, which is defined as $(a + b)/b$

That is,

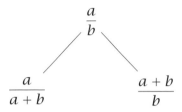

So the tree starts as follows:

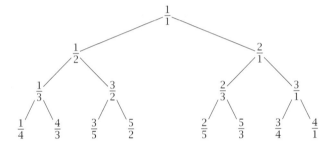

We construct a list of fractions from the nodes of the tree by traversing the tree from top to bottom, left to right to get

$$\frac{1}{1}, \frac{1}{2}, \frac{2}{1}, \frac{1}{3}, \frac{3}{2}, \frac{2}{3}, \frac{1}{4}, \frac{4}{3}, \ldots$$

It is not immediately apparent but, in fact, the process lists them all and the nice thing about it is that it does so by having each one of them appear precisely once. This is truly a 'proper' listing of the rationals.

Three elementary results combine to establish this and to consider them it is useful to make a definition.

Definition 7.1. A fraction a/b is called *reduced* if a and b have no prime factors in common. (Note that, by this definition, $\frac{1}{1}$ is reduced.) In fact, reduced is the same as coprime once we are past $\frac{1}{1}$.

Result 1. Every node is reduced.

To show this, suppose that a/b is a reduced node.

The left child is made up of a and $a + b$, and if these were not coprime there would exist k, c and d such that

$$a = ck \quad \text{and} \quad a + b = dk,$$

which means that

$$ck + b = dk \quad \text{and so } b = k(d - c)$$

and so b divides k, which means that it must divide $a = ck$, which is a contradiction.

Precisely the same argument establishes that the components of the right child $(a + b)/b$ must also be coprime.

Since $\frac{1}{1}$ is reduced we have an inductive proof of the result.

Result 2. Every positive reduced fraction appears as a node.

Consider the element a/b and define its *sum* as $a + b$. Assume the result holds for all fractions whose sum is k. We prove all fractions with the sum $k + 1$ must be on the tree.

Consider a fraction r/s such that $r + s$ is $k + 1$; it must be that $r \leqslant k$ and $s \leqslant k$. Further, $r \neq s$ since the fraction must be reduced, so either $r > s$ or $r < s$.

For $r > s$, evidently $r - s > 0$, and so $(r - s)/s > 0$. The sum of this fraction is $(r - s) + s = r \leqslant k$ and so $(r - s)/s$ must be a node by assumption; its right-hand child is $((r - s) + s)/s = r/s$ and so is a node.

For $r < s$, $s - r > 0$, and so $r/(s - r) > 0$. The sum of the fraction is $(r + s) - r = s \leqslant k$ so $r/(s - r)$ must be a node by assumption; its left-hand child is $r/((s - r) + r) = r/s$, which again is a node.

The starting condition $r + s = 2$ yields the unique positive fraction $\frac{1}{1}$, which is by definition a node, and the induction is once again complete.

Result 3. Every reduced fraction appears at most once.

We know that every fraction is reduced.

Suppose that the fraction j/k appears at least twice and call its two parents a/b and c/d. Obviously, $a/b \neq c/d$ because then j/k would be a left and a right child of the same node, but for integers a and b, $(a + b)/b > a/(a + b)$.

If j/k is a left child of both a/b and c/d, then $a/(a + b) = c/(c + d)$ and so $a/b = c/d$: the same argument ensures that both a/b and c/d cannot be a right child of the same parent. Therefore, one must be the left child of one parent and the right child of another.

Without loss of generality, we can assume j/k is the left child of a/b and the right child of c/d. Therefore, $j/k = a/(a + b)$ and $j/k = (c + d)/d$.

Because all fractions are reduced, this implies

$$j = a, \quad k = a + b, \quad j = c + d, \quad k = d,$$

and so

$$a = c + d \quad \text{and} \quad d = a + b$$

and $c + b = 0$. Since $b, c > 0$ we have the contradiction.

The result is then established.

The process generates the sequence of numerators 1, 1, 2, 1, 3, 2,..., which might be written as $b(n)$ and, since it can also be shown that the denominator of the nth fraction in the list equals the numerator of the $(n + 1)$th, the list of fractions is of the form

$$\frac{b(n)}{b(n + 1)}.$$

And we even have a nice recursive formula for the nth fraction on the list $a(n)$:

$$a(1) = \tfrac{1}{1},$$

$$a(n + 1) = \frac{1}{\lfloor a(n) \rfloor - (a(n) - \lfloor a(n) \rfloor) + 1}, \quad n \geqslant 1,$$

where '$\lfloor \cdot \rfloor$' is the floor function.

A Bigger Set

So, increasing dimension has no effect on countability, neither, as we have seen already, does increasing 'size': $\mathbb{Q}$, a set strictly containing $\mathbb{N}$, is countable. If we increase size again and consider the algebraic numbers which include the likes of $\sqrt{2}$ and all other numbers which are the roots of polynomial equations with integer coefficients, we get no further; a clever argument of Cantor showed that they are countable too. And so the adventure continued until intuition was for once shown to be a reliable guide: the whole real number system, $\mathbb{R}$, is not countable. Including the likes of π and e pushed matters too far, but if $\mathbb{Q}$ is countable and $\mathbb{R}$ not so, then what has been added—the *transcendental* numbers—must not be countable. In 1844 Joseph Liouville established an infinite class of transcendental numbers and in 1851 he managed to manufacture a particular number (now known as the Liouville number) which was provably transcendental, but finding such numbers proved to be extremely difficult. Now, with Cantor's result, there was the frustrating realization that 'almost all' numbers were one of these elusive transcendentals.

In fact, the proof that $\mathbb{R}$ is not countable is quite easy if the 'listability' of countable sets is utilized: more than this, $[0,1]$ is easily seen not to be listable. First, a small ambiguity is removed if we insist that finite decimals are represented by their infinite, 0.9 recurring, equivalent. For example, $0.284 = 0.283\,999\ldots$. Now suppose that $[0,1]$ is listable. But then consider the number $0.a_1a_2a_3\ldots$, which is formed by making a_1 anything other than the first decimal place of the first number in the list, a_2 anything other than the second decimal place of the second number in the list, etc.; by its construction, such a number is different from every number in the list and so the list cannot exhaust all numbers in the interval—and the required contradiction is in place.

With this result it is clear that any finite interval is uncountable, but to make things explicit we can do the following. If we restrict ourselves to the Euclidean world and so allow ourselves to accept the fifth postulate, we can see that two finite intervals

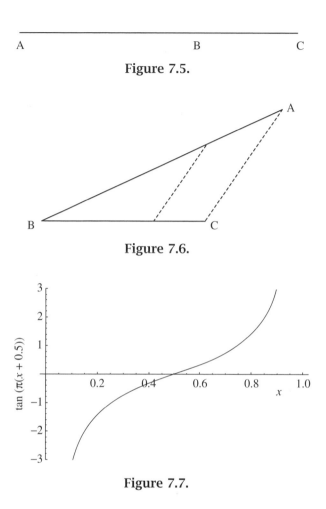

Figure 7.5.

Figure 7.6.

Figure 7.7.

of arbitrary length can be put into one-to-one correspondence by matching two line segments of their lengths, point for point.

Figure 7.5 shows two such segments, AB and BC, in line. Now fold the line, as shown in figure 7.6, to form an acute angle $\angle ABC$ and then join A and C. The Playfair reformulation of the fifth postulate guarantees that there is one and only one line through any point on AB which is parallel to AC; this line provides the one-to-one correspondence that we need.

All finite subintervals of $\mathbb{R}$ are therefore uncountable and we can establish a one-to-one correspondence of $[0, 1]$ with $\mathbb{R}$ using the function $f(x) = \tan(\pi(x + 0.5))$, as figure 7.7 suggests.

Cantor's Result

Returning to the earlier quotation of Cantor on page 72, what he had seen, and what he could not believe, was the resolution of a question articulated in an earlier, 1874 letter to Dedekind, in which Cantor asked:

> Can a surface (say a square that includes the boundary) be uniquely referred to a line (say a straight line segment that includes the end points) so that for every point on the surface there is a corresponding point of the line and, conversely, for every point of the line there is a corresponding point of the surface? I think that answering this question would be no easy job, despite the fact that the answer seems so clearly to be 'no' that proof appears almost unnecessary.

We have already seen on page 73 that $\mathbb{N}^n$ is countable, but that 1877 letter contained a proof which confounded the obvious by showing that there was a one-to-one correspondence of points on the interval $[0, 1]$ and points in n-dimensional space $\mathbb{R}^n$ or, equivalently, there is a one-to-one correspondence between $\mathbb{R}$ and $\mathbb{R}^n$: the whole is once again *not* necessarily greater than the part.

To set up the required one-to-one correspondence take any point (x, y) in the unit square, with the numbers given their infinite decimal form, and construct the decimal whose first decimal place is that of x, whose second decimal place is that of y, whose third decimal place the second of x, whose fourth decimal place the second of y, and so on. Such a number is unique in $[0, 1]$ and, conversely, any number in $[0, 1]$ can be disseminated uniquely into two numbers which are the x and y coordinates of a point in the unit square: the one-to-one correspondence is thereby established and the concept of dimension again brought into uncomfortable scrutiny. The same argument can be easily adapted to higher dimensions.

The whole theory is a firmament of fantastic results, most of which confound intuition, and some of which hit at the very heart of mathematical foundations, bringing with them genuine paradoxes which have shaken the assumed firm foundations of

the subject. One infamous example concerned the great logician Gottlob Frege, who had worked for a quarter of a century
on developing arithmetic from the basic logical foundations of
mathematics, as he defined them to be. The thesis was to occupy
two large volumes, with the first already published when Frege
received a letter from the august Bertrand Russell detailing his
own recent observations about set theory which were inspired
by Cantor's work. At the end of the second volume Frege had
added a footnote, which began:

> A scientist can hardly meet with anything more undesirable
> than to have the foundation give way just as the work is fin-
> ished. In this position I was put by a letter from Mr. Bertrand
> Russell as the work was nearly through the press.

Untold mathematicians and logicians have been deceived by the
implications of Cantor's work: arguments have raged and sides
have been taken. We will end the chapter on a positive note, with
the words of another German mathematician, David Hilbert (the
greatest of them all in that era): 'No one shall expel us from the
paradise that Cantor has created.'

Chapter 8

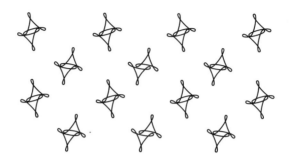

GAMOW–STERN ELEVATORS

For every complex problem, there is a solution that is simple, neat, and wrong.

H. L. Mencken

Gamow and Stern at Work

If we work on the middle floor of a building with one elevator and assume that floor usage is uniform, symmetry dictates that, if the elevator is not stationary on our floor, it will arrive at our floor with a probability $\frac{1}{2}$ of going up or down. Similar reasoning was used by George Gamow and Marvin Stern when, in 1956, they worked in a building of seven floors, with lowest floor numbered 1 and the highest numbered 7; Gamow's office was on the second floor and Stern's the sixth. Whenever Gamow decided to visit Stern the elevator almost always appeared as it was going down, and so the journey would be a frustrating 'down and then up'; a reciprocal situation was encountered by Stern as he tried to visit Gamow. With these numbers we can see from figure 8.1

Figure 8.1.

why they reasoned that, for Gamow, the probability is $\frac{1}{6}$ of the elevator arriving on the way up and $\frac{5}{6}$ of it arriving going down; Stern's situation was, of course, the opposite of this.

This is all very reasonable, but they wrote a little book of puzzles, called *Puzzle Math*, which was published in 1958. The elevator story is mentioned in the introduction and developed as a puzzle later in the book, with up-and-down elevators replaced by eastbound and westbound trains travelling between Chicago and Los Angeles. An argument based on a single train is given to conclude that an observer's proximity to each city determines the frequency with which eastbound and westbound trains are observed by him and concludes with the statement:

> If there are many trains travelling between Chicago and California, as is actually the case, the situation will, of course, remain the same, and the first train passing our city after any given time is still most likely to be an eastbound one.

The implication for the elevator problem is that, with any number of them in the building, the $\frac{1}{6}$ and $\frac{5}{6}$ probabilities would remain valid; actually, they do not. Whenever a mathematical statement contains the phrase 'of course' it lays itself open to particular scrutiny and in this case Donald Knuth provided

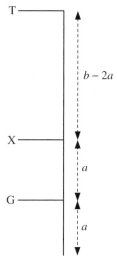

Figure 8.2.

that scrutiny and published his argument in a 1969 paper in the *Journal of Recreational Mathematics* (2:131–37). The argument demonstrates that, even though the elevators are moving independently of each other, Gamow's and Stern's probabilities change with their number—and, as that number increases, each probability approaches $\frac{1}{2}$.

Knuth's Argument

We will study the problem in the more general context of Gamow being a floors from the bottom of the building, which is taken to be b floors high, as in figure 8.2.

The case of the single elevator is easily established. The probability that the elevator will be descending as it arrives on Gamow's floor is $P_1 = (b - a)/b = 1 - p$ if we write $p = a/b$. In the Gamow-Stern case, $p = \frac{1}{6}$.

The subtlety begins to emerge when we consider a building with two elevators. We will consider the case $p \leqslant \frac{1}{2}$, thereby keeping G in the lower half of the building, supported by the reasoning that it is 'of course' clear that the alternative case is symmetrical to this one.

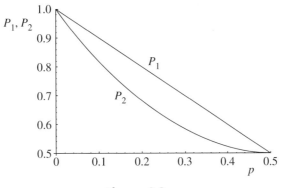

Figure 8.3.

A natural way to look at the problem is to argue that, if both elevators are above G, necessarily the next elevator will be travelling down; this occurs with a probability of $((b - a)/b)^2 = (1 - p)^2$. The only other way that the next elevator could be travelling down is when one elevator is above and the other below G; this occurs with a probability

$$\frac{a}{b} \times \frac{b - a}{b} + \frac{b - a}{b} \times \frac{a}{b} = 2p(1 - p).$$

To calculate the contribution of this to the full probability, we need to multiply by the probability that the one above is close enough to compete successfully with the one below, that is, it is between G and X; this probability is

$$\frac{a}{b - a} = \frac{p}{1 - p};$$

then we must multiply by $\frac{1}{2}$, as both elevators are in equal competition, to give the total probability of the next elevator going down as

$$P_2 = (1 - p)^2 + 2p(1 - p) \times \frac{p}{1 - p} \times \frac{1}{2}$$
$$= 1 - 2p + p^2 + p^2 = 1 - 2p + 2p^2.$$

Figure 8.3 is a plot of P_1 and P_2 against p and we can see that $P_1 > P_2$, apart from the extreme values.

The generalization of the result to n elevators can be pursued in several ways. The one we will adopt is to notice that the two results that we have so far can be rewritten as

$$P_1 = 1 - p = \tfrac{1}{2} + \tfrac{1}{2}(1 - 2p),$$
$$P_2 = 1 - 2p + 2p^2 = \tfrac{1}{2} + \tfrac{1}{2}(1 - 2p)^2,$$

which could suggest that $P_n = \tfrac{1}{2} + \tfrac{1}{2}(1 - 2p)^n$ for all positive integers n.

In this new form, the arguments used to arrive at P_1 and P_2 change to the following.

With one elevator, the elevator could be above X and so would assuredly be coming down; this occurs with a probability

$$\frac{b - 2a}{b} = 1 - 2p.$$

Alternatively, it could be below X, which occurs with a probability $1 - (1 - 2p)$, and there is an even chance of it then being above G or below it and hence going up or down, so this probability becomes $\tfrac{1}{2}(1 - (1 - 2p))$ and so

$$P_1 = (1 - 2p) + \tfrac{1}{2}(1 - (1 - 2p)) = \tfrac{1}{2} + \tfrac{1}{2}(1 - 2p).$$

Similarly, with two elevators, both of them could be above X, with a probability of

$$\left(\frac{b - 2a}{b}\right)^2 = (1 - 2p)^2$$

or at least one of them will be below X, with a probability $1 - (1 - 2p)^2$, in which case there is an even chance of the next elevator being above X, which gives

$$P_2 = (1 - 2p)^2 + \tfrac{1}{2}(1 - (1 - 2p)^2) = \tfrac{1}{2} + \tfrac{1}{2}(1 - 2p)^2.$$

With this in mind, the general case can be argued in precisely the same way, with n replacing 2, to give

$$P_n = (1 - 2p)^n + \tfrac{1}{2}(1 - (1 - 2p)^n) = \tfrac{1}{2} + \tfrac{1}{2}(1 - 2p)^n.$$

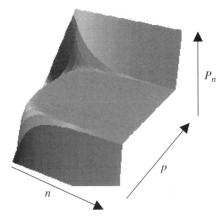

Figure 8.4.

The formula for P_n can easily be modified to allow p to take values greater than $\frac{1}{2}$, and it becomes

$$P_n = \tfrac{1}{2} + \tfrac{1}{2}(1 - 2p)|(1 - 2p)|^{n-1},$$

the plot of which is shown in figure 8.4, with n taken as a continuous variable.

With $n = 1$, we can see at the rear of the graph the straight line $P_1 = 1 - p$ and as n increases, P_n becomes the plateau at height $\frac{1}{2}$ for all values of p.

We will leave the elevator paradox at this point, but the reader may wish to pursue matters further and could do so, for example, by reading A. Wuffle's article 'The pure theory of elevators' (1982, *Mathematics Magazine* 55(1):30–37) or, unsurprisingly, Martin Gardner with his chapter discussing elevators in *Knotted Doughnuts and Other Mathematical Entertainments*. A more gentle pursuit would be afforded by watching the CBS television series *NUMB3RS*, which we have already mentioned in chapter 6; the 14 December 2007 episode 'Chinese Box' has Alan and Charlie suffering from its effect.

Chapter 9

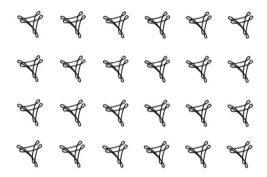

THE TOSS OF A COIN

Constant repetition carries conviction.

Robert Collier

A Modern Matter

If the reader was asked to judge whether the following 1679 bits of binary data is random or could possibly contain a message, the answer would most probably be the latter: the eye discerns some sort of 'lack of randomness' with those sequences of 0s surely too long for mere chance to have created them. Think of tossing a coin 1679 times and writing a 0 if a head appears uppermost and a 1 otherwise: such long runs of heads would surely cause us to suspect the coin's fairness. If the reader agrees with this, he or she will be acting precisely as the Communication with Extra Terrestrial Intelligence (CETI) group would wish. This

Arecibo Message was beamed into space by them in the direction of the 25 000 light years distant globular star cluster M13 to commemorate the remodelling of the Arecibo radio telescope in 1974:

```
00000010101010000000000010100000101000000010010001000100010001001
01100101010101010101010010010000000000000000000000000000000000000
00011000000000000000000011010000000000000000000110100000000000000
00000001010100000000000000000011111000000000000000000000000000000
00000110000111000110000110001000000000000001100100001101000110000
01100001101011111011111101111101111100000000000000000000000000001
00000000000000000010000000000000000000000000000001000000000000000
00111111000000000000011111000000000000000000000000001100001100001
11000110001000000010000000001000011010000110001110011010101111100
11111101111101111100000000000000000000000000001000000110000000001
00000000000110000000000000000010000011000000000001111110000011000
00011111000000000001100000000000010000000010000000010000010000000
00110000000100000001100001100000010000000000011000100001100000000
00000000011001100000000000001100010000110000000001100001100000000
01000000010000001000000001000001000000011000000001000100000000000
11000000001000100000000010000000010000010000000100000001000000000
10000000000001100000000011000000001100000000010001110101100000000
00000010000000100000000000000010000011110000000000001000010111
01001011011000000100111001001111111011100001110000011011100000
00001010000011101100100000010100000111111001000000101000001100
00001000001101100000000000000000000000000000000000000111000001000
00000000000011101010001010101010100111000000000101010100000000000
00000001010000000000000001111100000000000000000011111111100000000
00001110000000111000000000011000000000001100000001101000000000001
01100000110011000000011001100001000101000001010001000010001001
00010010001000000001000101000100000000000010000100001000000000
00010000000001000000000000001001010000000000011110011111010011
1100
```

Any intelligent extraterrestrial life (or reader!) would soon factor 1679 into primes as 23×73 to suggest a rectangular grid with 23 rows and 73 columns or 73 rows and 23 columns, filling in the 1 squares and leaving the 0 squares blank. Figure 9.1 shows the result of the former decomposition, which might dampen the spirit, but figure 9.2 shows the latter as one which contains (albeit somewhat elusive) information and the reader might wish to pursue the matter further to find out just what that message is. Having made the point that intuition has guided us correctly,

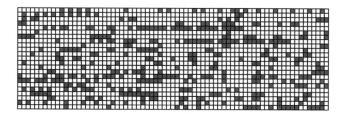

Figure 9.1.

Figure 9.2.

we will not pursue the details of the 1999 *Encounter 2001 Message* to a giant star cluster in Hercules—with its error-protected ~400 000 binary digits!

More modestly, consider the two 200 bits of binary data:

01000000011100011000011101000011000001011010000000001101011110000
11001000101101101010000101010011101101110000110000110101110000
11110001000111110001001001101101111110111001000001111100001001
01001001010011

and

1011000101011010001011110000100001010100110010001110011010100111
0110101111000111101001101110011101101011110101011011010100110000
1001100110010010010101000011110101110000101011110011010101111000
10010011000100

The test is sterner but one of the sets is random, the other not
so: which of the two might contain information? Using the same
intuition and analogy, the first data set would have us believe
that the coin came up heads eight times in a row whereas the
second at most four times in a row: the second is random data,
the first not, then. Not so.

Associating 0s and 1s with heads and tails, the reader who
chose the wrong sequence might seek comfort in the words of
the late, celebrated, former Harvard statistician Fred Mosteller;
he is quoted by Victor Cohn in his book *News & Numbers: A
Guide to Reporting Statistical Claims and Controversies in Health
and Other Fields* (Iowa State University Press, 1989) as follows:

> If you toss a coin repeatedly in a college class and after
> each toss ask the class if there is anything suspicious going
> on, 'hands suddenly go up all over the room' after the fifth
> head or tail in a row. There happens to be only 1 chance in
> 16 (0.0625)—not far from 0.05, or 5 chances in 100—that 5
> heads or tails in a row will show up in five tosses, 'so there
> is some empirical evidence that the rarity of events in the
> neighbourhood of 0.05 begins to set people's teeth on edge'.

If we continue with the model of coin tossing, we shall see that
the behaviour of sequences of heads (or tails) is rather different
to what one might reasonably imagine.

An Enlightened Approach

The investigation is not a new one and we will first look at
the approach of the eighteenth-century French mathematician
Abraham De Moivre, of whom Isaac Todhunter said of probability theory that it 'owes more to [De Moivre] than any other
mathematician, with the single exception of Laplace.'

The opinion appeared in his vastly influential book *A History of
the Mathematical Theory of Probability from the Time of Pascal*

to that of Laplace (1865, reprinted 1965): in the same volume he dated the beginnings of the modern theory of probability to 1654, born of a late summer exchange of letters between Pierre Fermat and Blaise Pascal. This formative interchange was partly inspired by *The Problem of Points* put to Pascal by the notable gambling aristocrat Antoine Gombaud, chevalier de Méré, sieur de Bassay. De Méré had asked another far-from-new question, which is naturally phrased in terms of the toss of a coin:

> Suppose two players A and B stake equal money on being the first to win n points in a game in which the winner of each point is decided by the toss of a fair coin, heads for A and tails for B. If such a game is interrupted when A still lacks a points and B lacks b points, how should the total stake be divided between them?

After several special cases had been dealt with, the general solution to the problem was eventually decided upon, which brought to the mathematical world the first association of the name of Pascal with the famous numerical triangle. For interest (and without proof) the solution is that, from the $(a + b - 1)$th row of Pascal's triangle, player A should receive

$$\frac{\text{Sum of the first } b \text{ entries}}{\text{Sum of the entire row}} \times n$$

$$= \frac{1}{2^{a+b-1}} \left(\binom{a+b-1}{0} + \binom{a+b-1}{1} + \cdots \right.$$
$$\left. + \binom{a+b-1}{b-2} + \binom{a+b-1}{b-1} \right) \times n$$

and player B should receive

$$\frac{\text{Sum of the remaining entries}}{\text{Sum of the entire row}} \times n$$

$$= \frac{1}{2^{a+b-1}} \left(\binom{a+b-1}{b} + \binom{a+b-1}{b+1} + \cdots \right.$$
$$\left. + \binom{a+b-1}{a+b-2} + \binom{a+b-1}{a+b-1} \right) \times n.$$

For example, if $n = 10$ and the play stops when A has accumulated eight points and B seven points, we have

$$a = 10 - 8 = 2, \quad b = 10 - 7 = 3, \quad a + b - 1 = 4.$$

Thus, player A should receive

$$\frac{1}{2^4} \left(\binom{4}{0} + \binom{4}{1} + \binom{4}{2} \right) \times 10 = 6.875$$

and player B should receive

$$\frac{1}{2^4} \left(\binom{4}{3} + \binom{4}{4} \right) \times 10 = 3.125.$$

Another luminary whose light shone with fierce brightness in the probabilistic firmament (and equally pretty much everywhere else) was the Swiss mathematician Jacob Bernoulli, whose fundamental result connected with summing elements of the rows of Pascal's triangle will occupy us a little later (and we will meet him again in chapter 11), but now we will concern ourselves with De Moivre. What started out as a submission to the Royal Society in 1711 was developed into the volume *The Doctrine of Chances* of 1718, reprinted in 1738 and again in a third edition in 1756. The title page of this last edition has it that its content is 'Fuller, Clearer, and more Correct than the Former'.

We will see that the considerable problems associated with ordinary calculation continued to be costly, but it is from this edition that we will discuss his investigation into repetition when a coin is tossed. The majority of the book is devoted to the statement of and answers to a series of problems, with the last one of the 'Chance' section (immediately proceeding the second part of the book, which is devoted to 'A Treatise of Annuities on Lives') the following, which we have modernized slightly.

Problem LXXIV

To find the probability of throwing a chance assigned a given number of times without intermission, in any given number of trials (that is, the probability of achieving a given number of repetitions in a given number of trials).

Solution. Let the probability of throwing the chance in any one trial be represented by $a/(a + b)$ and the probability of the contrary by $b/(a + b)$. Suppose n to represent the number of trials given, and p the number of times that the chance is to come up without intermission; then supposing $b/(a + b) = x$, take the quotient of unity divided by

$$1 - x - axx - aax^3 - a^3x^4 - a^4x^5 - \cdots - a^{k-1}x^k,$$

and having taken as many terms of the series resulting from that division as there are units in $n - k + 1$, multiply the sum of the whole by a^kx^k/b^k, or by $a^k/(a + b)^k$, and that product will express the probability required.

Example 1. Let it be required to throw the chance assigned three times together, in 10 trials, when a and b are in a ratio of equality, otherwise when each of them is equal to unity; then having divided 1 by $1 - x - xx - x^3$, the quotient continued to so many terms as there are units in $n - k + 1$, that is, in this case to $10 - 3 + 1 = 8$, will be

$$1 + x + 2xx + 4x^3 + 7x^4 + 13x^5 + 24x^6 + 44x^7.$$

Where x being interpreted by $b/(a + b)$, that is, in this case by $\frac{1}{2}$, the series will become

$$1 + \tfrac{1}{2} + \tfrac{2}{4} + \tfrac{4}{8} + \langle\tfrac{7}{8}\rangle + \tfrac{7}{16} + \tfrac{13}{32} + \tfrac{24}{64} + \tfrac{44}{128},$$

of which the sum is $\frac{520}{128} = \frac{65}{16}$, and this being multiplied by a^kx^k/b^k, that is, in this case by $\frac{1}{8}$, the product will be $\frac{65}{128}$, and therefore 'tis something more than an equal chance that the chance assigned will be thrown three times together some time in 10 trials, the odds for it being 65 to 63.

In modern terms, the odds of there being a sequence of at least three heads in ten tosses of a fair coin are 65:63. First, notice any author's nightmare, the typo '$\frac{7}{8}$'. To the modern eye this is all very confusing and we will try to remove the contortions by using modern notation. Consider the expression

$$E = \frac{1}{1 - x - axx - aax^3 - a^3x^4 - a^4x^5 - \cdots - a^{k-1}x^k} \times \frac{a^k x^k}{b^k}$$

$$= \frac{1}{1 - x - ax^2 - a^2x^3 - a^3x^4 - a^4x^5 - \cdots - a^{k-1}x^k} \times \frac{a^k x^k}{b^k}$$

and, using the formula for the sum of a geometric series,

$$E = \frac{1}{1 - x(1 - (ax)^k)/(1 - ax)} \times \frac{a^k x^k}{b^k}$$

$$= \frac{1 - ax}{1 - ax - x + a^k x^{k+1}} \times \frac{a^k x^k}{b^k}.$$

Our principal interest lies with a fair coin, for which $a = b = 1$, and the expression becomes

$$E = \frac{1 - x}{1 - 2x + x^{k+1}} \times x^k = \frac{x^k(1 - x)}{1 - 2x + x^{k+1}}.$$

If we write $P(n, k)$ as the probability that, in n tosses of a fair coin, a sequence of heads of length at least k appears, we have the compact pseudo-expression

$$P(n, k) = \text{Expand}\left[\frac{x^k(1 - x)}{1 - 2x + x^{k+1}} : \text{ up to } x^n, \; x = \tfrac{1}{2}\right].$$

His example asks for the value of $P(10, 3)$ and, as he says,

$$E = \frac{x^3(1 - x)}{1 - 2x + x^4}$$

$$= x^3(1 + x + 2x^2 + 4x^3 + 7x^4 + 13x^5 + 24x^6 + 44x^7 + O(x^8)),$$

and evaluating the significant part of this at $x = \tfrac{1}{2}$ gives

$$P(10, 3) = \tfrac{1}{8}(1 + \tfrac{1}{2} + \tfrac{2}{4} + \tfrac{4}{8} + \tfrac{7}{16} + \tfrac{13}{32} + \tfrac{24}{64} + \tfrac{44}{128}) = \tfrac{65}{128} \approx 0.508.$$

His formula, which is a type of generating function, is deeply mysterious and offered without any justification, although he does continue 'to consider of Expedients to make the Calculation more easy', illustrating his point with reference to $P(21, 4)$. The essence of the computational difficulty is in the expansion of $1/(1 - 2x + x^5)$, which to the required degree of accuracy is given as

$$1 + 2x + 4x^2 + 8x^3 + 16x^4 + 31x^5 + 60x^6 + 116x^7 + 224x^8$$
$$+ 432x^9 + 833x^{10} + 1606x^{11} + 3096x^{12} + 5968x^{13}$$
$$+ \langle 11\,494x^{14} + 22\,155x^{15} + 42\,704x^{16} + 82\,312x^{17} \rangle$$

with our angled brackets indicating a calculative error. The accurate expression, again produced by a computer in the blink of an eye, is

$$1 + 2x + 4x^2 + 8x^3 + 16x^4 + 31x^5 + 60x^6 + 116x^7 + 224x^8$$
$$+ 432x^9 + 833x^{10} + 1606x^{11} + 3096x^{12} + 5968x^{13}$$
$$+ 11\,504x^{14} + 22\,175x^{15} + 42\,744x^{16} + 82\,392x^{17}.$$

A single error in calculation at the x^{14} term has compounded itself throughout the remainder of the expression by his expedient use of a mysterious recursion:

> Now although these Terms may seem at first sight to be acquired by very great labour, yet if we consider what has been explained before concerning the nature of a recurring Series, we shall find that each Coefficient of the Series is generated from the double of the last, subtracting once the Coefficient of that Term which stands 5 places from the last inclusive; so that for instance if we wanted one Term more, considering that the last Coefficient is 82 312, and that the Coefficient of that Term which stands five places from the last inclusive is 5968, then the Coefficient required will be twice 82 312, wanting once 5968, which will make it 158 656, so that the Term following the last will be $158\,656x^{18}$.

The prescription works but one is often left to wonder what these great mathematicians would have done with modern computing power, which gives the instant answer $P(21, 4) = 0.497$.

A Modern Approach

With a computer, De Moivre's method provides an extremely easy, yet puzzling, way of computing the probabilities but, for the sake of mathematical probity, we will use more rigorous methods to develop a recurrence relation in n for $P(n, k)$.

That sequence of k heads might occur within the first $n - 1$ tosses or it might not. If it does, it does so with a probability of $P(n - 1, k)$; if it does not, the final coin toss must be included in the sequence and probability that this happens can be calculated in the following way: there must have been no sequence of k heads in the first $n - k$ tosses and the $(n - k)$th toss must be a tail, which is followed by the sequence of k heads. This gives rise to the probability

P(No sequence of k heads in first $n - k$ tosses)

$\quad \times P$(A sequence of k heads which includes the nth toss)

$$= [1 - P(n - k, k)] \times \tfrac{1}{2}(\tfrac{1}{2})^k = [1 - P(n - k, k)] \times (\tfrac{1}{2})^{k+1}.$$

We then have

$P(n, k)$

$\quad = P$(That the sequence of heads occurs within the first tosses)

$\qquad + P$(That the sequence of heads includes the final toss)

and so

$$P(n, k) = P(n - 1, k) + [1 - P(n - k, k)] \times (\tfrac{1}{2})^{k+1}.$$

Evidently, we also have the boundary conditions

$$P(0, k) = P(1, k) = P(2, k) = \cdots = P(k - 1, k) = 0$$

and

$$P(k, k) = (\tfrac{1}{2})^k$$

Using these we can generate the probabilities recursively and so arrive at table 9.1.

Table 9.1. $P(n, k)$.

			k			
n	3	4	5	6	7	8
10	0.508	0.245	0.109	0.047	0.020	0.008
20	0.787	0.478	0.250	0.122	0.058	0.027
30	0.908	0.639	0.368	0.192	0.095	0.046
40	0.960	0.750	0.468	0.256	0.131	0.065
50	0.983	0.827	0.552	0.315	0.165	0.084
100	1	0.973	0.810	0.546	0.318	0.170
200	1	1	0.966	0.801	0.544	0.320

We see De Moivre's value for $P(10, 3)$ appearing; also, we see that the event that the eight consecutive heads appear in 200 tosses of the coin, $P(200, 8)$, is not negligible.

Heads or Tails

Of course, it is the fact that a 'long' repetition appears that causes surprise, whether it be wholly heads or wholly tails. For a complete resolution of the situation we really need to be able to find what we will write as $Q(n, k)$, the probability of a sequence of heads or tails of length k appearing in n tosses of a fair coin.

To achieve this we could revert to the previous method, but there is another, neater approach.

A run of heads or tails continues with the appearance of the pairs HH or TT and ends with the appearance of the opposite outcome and so with the sequence HT or TH—each of the four possibilities appearing with equal likelihood. Represent the event that a consecutive pair are the same by S and that they are different by D and consider the sequence of Ss and Ds so generated. For example, consider the sequence of Hs and Ts

HHHTTHTHHTTTTHH.

This generates the associated sequence of Ss and Ds

H	H	H	T	T	H	T	H	H	T	T	T	T	H	H
	S	S	D	S	D	D	D	S	D	S	S	S	D	S

Table 9.2. $Q(n,k)$.

n	3	4	5	6	7	8
10	0.826	0.465	0.217	0.094	0.039	0.016
20	0.950	0.720	0.458	0.237	0.115	0.054
30	0.994	0.879	0.625	0.357	0.185	0.092
40	0.999	0.948	0.741	0.459	0.250	0.128
50	1	0.981	0.821	0.544	0.309	0.162
100	1	1	0.972	0.807	0.542	0.315
200	1	1	0.999	0.965	0.799	0.542

The three consecutive Ss guarantee four consecutive Ts. If we recognize that the length of this new sequence is precisely one less than the original, our problem of what is

$Q(n,k)$ = the probability of a sequence of H or T of
length k appearing in n tosses of a fair coin

reduces to what is

$Q(n,k)$ = the probability of a sequence of Ss of
length $k-1$ appearing in the generated
sequence of $n-1$ Ss and Ds.

This means that we have $Q(n,k) = P(n-1, k-1)$ and using this we can compile table 9.2.

Now we see that getting a run of length 8 in 200 tosses of the coin is more likely than not! It is not unusual for instructors to bring the phenomenon to the attention of students by asking half the class to toss a fair coin (say) 100 times and to record the outcome on a named sheet of paper and the other half to make up the data, as if they has tossed the coin. Not many would be brave enough to make up a sequence of either H or T of length 5 or greater, yet the chance of this happening we can see is 97%: the instructor can hand back the sheets with just a 3% chance of being wrong in each case. Reverting to the Arecibo data, there is a sequence of 0s of length 37 and $Q(1679, 37) = 1.2 \times 10^{-8}$, so intuition is not always confounded!

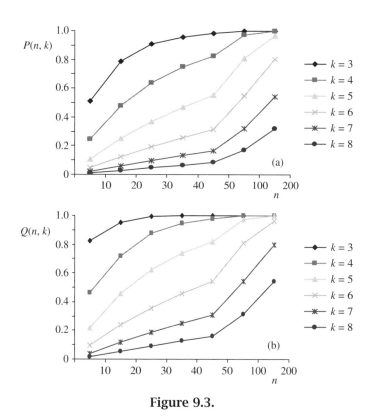

Figure 9.3.

Figure 9.3 indicates the trends of $P(n,k)$ and $Q(n,k)$ using the data of the previous tables.

Asymptotic Behaviour

We can ask the question: how long do we expect to wait for a run of n consecutive heads? If we write E_n for the expected number of tosses to achieve a run of n consecutive heads, we can form a recurrence relation in the following way.

We can achieve the n consecutive heads by realizing $n-1$ consecutive heads and then another head or by failing by getting a tail at the final toss of the coin, and then starting over again. This brings about the relation

$$E_n = \tfrac{1}{2}(E_{n-1} + 1) + \tfrac{1}{2}(E_{n-1} + 1 + E_n)$$

and this easily reduces to $E_n = 2E_{n-1} + 2$.

There is a standard theory that deals with what is character-ized as *a linear nonhomogeneous recurrence relation* but we will assume an answer of $E_n = 2^{n+1} - 2$ and show it to be so using induction.

The relation tells us that $E_1 = 2^{1+1} - 2 = 2 = 1/(\frac{1}{2})$ and from the theory of the geometric random variable, or using common sense, this is clearly true. Now assume it true for $n = k$ and so $E_k = 2^{k+1} - 2$. Then $E_{k+1} = 2E_k + 2 = 2(2^{k+1} - 2) + 2 = 2^{k+2} - 2$ and the induction step is successfully taken.

Much more challenging is the question of the expected maxi-mal length of a run of heads in n tosses of a coin. If R_n is defined as the random variable, the maximum length of a run of heads in n tosses of a coin, then Mark Schilling et al. (L. Gordon, M. F. Schilling and M. S. Waterman, 1986, An extreme value theory for long head runs, *Probability Theory and Related Fields* 72:279–87, and M. F. Schilling, 1990, The longest run of heads, *The College Mathematics Journal* 21(3):196–207) have shown that

$$E[R_n] = \log_2 \frac{n}{2} + \frac{y}{\ln 2} - \frac{1}{2} + r_1(n) + \varepsilon_1(n),$$

$$V[R_n] = \frac{\pi^2}{6} \times \frac{1}{(\ln 2)^2} + \frac{1}{12} + r_2(n) + \varepsilon_2(n),$$

where $y = 0.577\ldots$ is Euler's constant (the reader can find out about this mysterious number by consulting an earlier book by this author, *Gamma: Exploring Euler's Constant*, Princeton Uni-versity Press, 2003). Also, $|r_1(n)| \leqslant 0.000\,016$ and $|r_2(n)| \leqslant 0.000\,06$ for all n and $\varepsilon_1(n), \varepsilon_2(n) \to 0$ as $n \to \infty$.

So, as n increases we have the estimate that

$$E[R_n] = \log_2 n - \tfrac{2}{3}.$$

And since we have the remarkable fact that the variance is virtually constant at

$$V[R_n] \approx \frac{\pi^2}{6} \times \frac{1}{(\ln 2)^2} + \frac{1}{12} \approx 3.507$$

and the standard deviation at 1.873, we can see that the esti-mated length of the longest run is very accurate.

Schilling points out in the later article that

> The variety of potential applications of runs theory is virtu-
> ally boundless. Some of the more intriguing include hand-
> writing analysis by means of digitized scanning, hydrologic
> runs (floods and droughts) and studies of the pattern of
> capture of prey species.

Chapter 10

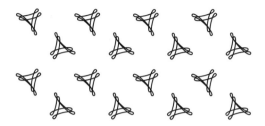

WILD-CARD POKER

These results, which are partly combinatorial and partly real mathematics...

A. Joseph

Poker Hands

Since its invention somewhere in the Louisiana Territory around 1800, poker has evolved into a vastly complex contest of skill and chance. Acknowledged experts spread across the globe, even the late and acclaimed bridge expert, Terence Reese, coauthored a book entitled *Poker: Game of Skill*. The game exists in many variants, which add novelty and subtlety to the standard rules, but whatever the details of the particular variant there is a ranking of the winning hands. To begin with we list the standard hands, in the order of their rankings (see also table 10.1):

Straight flush: all five cards in a consecutive sequence in the same suit.

Four of a kind: four cards of one denomination and a fifth odd card.

Full house: three cards of one denomination and two of another.

Table 10.1. Standard poker-hand rankings.

Royal flush	10 ♠	J ♠	Q ♠	K ♠	A ♠
Straight flush	3 ♣	4 ♣	5 ♣	6 ♣	7 ♣
Four of a kind	K ♥	K ♦	K ♣	K ♠	4 ♥
Full house	10 ♦	10 ♥	10 ♠	A ♣	A ♠
Flush	8 ♠	Q ♠	2 ♠	5 ♠	6 ♠
Straight	6 ♥	7 ♣	8 ♠	9 ♣	10 ♦
Three of a kind	7 ♠	7 ♥	7 ♦	J ♠	A ♠
Two pairs	A ♣	A ♥	6 ♠	6 ♣	J ♠
One pair	J ♠	J ♦	2 ♥	5 ♣	9 ♠

Flush: all five cards in the same suit but not consecutive.

Straight: all five cards consecutive but not of the same suit.

Three of a kind: three cards of the same denomination and two odd cards.

Two pairs: two cards of the same denomination twice and an odd card.

One pair: two cards of the same denomination.

Card high: anything else.

These rankings are decided by the likelihood of each hand being dealt to a player. To investigate the idea we will take the simplest form of poker, in which five cards are dealt to each player from a standard pack of fifty-two cards. The whole analysis is simply a counting exercise, which is made easier by the use of the standard combination formula

$$\binom{n}{r} = \frac{n!}{r!(n-r)!}$$

for the number of ways of choosing r objects from n without regard to order.

Creating the Hierarchy

Straight flush

The whole hand is determined by the smallest card, which can be any of ace, 2, 3, up to 10, and each of these cards has four possible suits, which means that there are $10 \times 4 = 40$ possible straight flushes.

Four of a kind

We can build four of a kind by first picking the card to match with the three others, and then pick the odd card out. Since we can pick the first card in 13 ways (without regard to suit), leaving 48 cards remaining after its removal and that of its three matching cards, we have $13 \times 48 = 624$ possible hands.

Full house

We pick the three of a kind, by picking the card to match with the two others, which can be done in 13 ways, and then multiplying by the number of ways the three suits can be chosen from the four possible, that is, $\binom{4}{3}$ ways: this gives us $13 \times \binom{4}{3}$ possibilities. The first of the two remaining cards can then be picked in 12 ways and again we have to count the number of ways the two suits of the pair can be chosen; this is $\binom{4}{2}$ ways, which gives us $12 \times \binom{4}{2}$ possibilities. The total number of possibilities is then

$$13 \times \binom{4}{3} \times 12 \times \binom{4}{2} = 3744.$$

Flush

There are four ways of choosing the suit and having done so there are $\binom{13}{5}$ ways of choosing the five cards and so $4 \times \binom{13}{5}$ possibilities. From this total we have to subtract the 40 straight flushes to finish with the total number of possibilities $4 \times \binom{13}{5} - 40 = 5108$.

Straight

We will take the case of 'round the corner' straights (e.g. JQKA2) not being allowed. The straight is determined by its lowest card, just as with the straight flush, and we have 40 possibilities for this. The remaining four cards have their denomination determined but each can come from any of the four suits, which means that we have 40×4^4 possibilities, 40 of which will be straight flushes. The final total is then $40 \times 4^4 - 40 = 10\,200$.

Three of a kind

As with the full house, the three cards can be chosen in $13 \times \binom{4}{3}$ ways and since we may not choose the fourth of the same denomination we have 48 cards left from which to choose the remaining two cards; this can be done in $\binom{48}{2}$ ways. This total will include the full houses and so we must subtract the 3744 of those to leave $13 \times \binom{4}{3} \times \binom{48}{2} - 3744 = 54\,912$ different hands.

Two pairs

To get two pairs, we first choose one pair by picking the first card, then a second card to match it from the remaining three suits, which can be done in $13 \times \binom{4}{2}$ ways. Then, we pick the second pair, making sure we do not match the first pair chosen, by picking one of the 48 remaining cards. We match it by picking another suit, which can be done in $12 \times \binom{4}{2}$ ways. The last card we pick must not match either of the first pairs, and so we are left with 44 cards from which to choose. Lastly, the two pairs can appear in either order and so we must divide this total by 2 to finish with $13 \times \binom{4}{2} \times 12 \times \binom{4}{2} \times 44 \times \frac{1}{2} = 123\,552$ ways.

One pair

To get a single pair and nothing else, we first pick one pair in $13 \times \binom{4}{2}$ ways and then pick the next three cards, making sure that there is no match. To do this we successively pick a card and then discard all cards of that denomination and this can be done in $48 \times 44 \times 40$ ways. The total number of ways of picking two pairs is then $13 \times \binom{4}{2} \times 48 \times 44 \times 40 = 1\,098\,240$.

Odd card

The remaining alternative is that the hand contains five odd cards and we can calculate the number of ways of achieving this by subtracting the total of all of the above from the total number of possible hands to get

$$\binom{52}{5} - 624 - 3744 - 5108 - 10\,200 - 54\,912 - 123\,552 = 1\,302\,540$$

ways.

Table 10.2. Natural frequencies of poker hands.

Hand	Frequency	Probability	Odds
Straight flush	40	0.000 015 4	64 973 : 1
Four of a kind	624	0.000 240	4 164 : 1
Full house	3 744	0.001 44	693 : 1
Flush	5 108	0.001 97	508 : 1
Straight	10 200	0.003 92	254 : 1
Three of a kind	54 912	0.021 1	46 : 1
Two pairs	123 552	0.047 5	20 : 1
One pair	1 098 240	0.423	1.37 : 1
Odd card	1 302 540	0.501	0.995 : 1

This is all summarized in table 10.2, where the probability is computed simply by dividing each frequency by $\binom{52}{5}$ and the hands are listed in descending order of value, measured by the probability of them occurring. The odds are calculated with the understanding that, if the odds of an event happening are given as $a : b$, it means that the probability of that event happening is $b/(a + b)$. In our case, $b = 1$ and so $p = 1/(a + 1)$, which makes $a = 1/p - 1$.

Wild-card poker

So far we have nothing more than a rather messy set of calculations which justify the standard hand ranking. Now suppose that we consider one of the most common variants of the standard game: the introduction of wild cards. Again, there are many possibilities and we will choose to add a single joker to the pack and allow it to count for any card. We need to recompute the probabilities, which is a little more delicate than before.

Five of a kind

This new possibility occurs precisely when the hand is four of a kind together with the joker. This occurs in thirteen possible ways.

Straight flush

Care must be taken with the card the joker replaces. The hands with lowest card an ace to a 9 all behave in the same way, taking the ace as typical and using '$*$' for joker, we have

$$\begin{pmatrix} A234* \\ A23*5 \\ A2*45 \\ A*345 \end{pmatrix}$$

as possible straight flushes in any of the four suits. The fifth possibility of $*2345$ would be counted as the superior hand $2345*$ or 23456. This rollover of the final possibility does not take place with the final possibility, where the lowest card is a 10; this will simply count as the fifth way of achieving that hand (recall that 'round the corner' straights are not permitted). The count becomes

$$4 \times (9 \times 4 + 5) = 164$$

possible ways. Add this to the 40 without a joker and the grand total becomes 204 ways.

Four of a kind

We already know that there are 624 ways without a joker; with it we need three of a kind, the joker and an odd card. This means that there are

$$624 + \binom{13}{1} \times \binom{4}{3} \times 1 \times \binom{48}{1} = 3120$$

ways.

Full house

Now we have to start being very careful. We can achieve the hand naturally in 3744 ways or with the joker added to two pairs; notice that we would not add the joker to three of a kind as that would make up the higher hand of four of a kind. The count is then

$$3744 + \binom{13}{2} \times \binom{4}{2} \times \binom{4}{2} \times 1 = 3744 \times 6 \times 6 = 6552.$$

Flush

The reasoning is much the same as before, we simply need four cards of the same suit. This is achieved in $4 \times \binom{13}{4} = 2860$ ways. Now subtract the straight flushes to get $2860 - 164 = 2696$ ways and then add in the natural ways to get $2696 + 5108 = 7804$ ways.

Straight

Once again we need to treat the cases where the lowest card is ace to 9 as one and the highest straight separately. For the nine lowest straights there are each 4^4 ways of choosing the four cards needed and they can each come from any of the four suits, with nine possible lowest cards we have $9 \times 4 \times 4^4$ possibilities. There will be $4^4 \times 5$ possibilities for that. Now we have to subtract the straight flushes to get $9 \times 4 \times 4^4 + 4^4 \times 5 - 164 = 10\,332$ ways. Finally, adding in the natural possibilities gives a grand total of $10\,332 + 10\,200 = 20\,532$.

Three of a kind

To add to the $54\,912$ natural possibilities, we need one pair, the joker and two odd cards, to get

$$54\,912 + \binom{13}{1} \times \binom{4}{2} \times 1 \times \binom{12}{2} \times \binom{4}{1} \times \binom{4}{1} = 137\,280$$

ways.

Two pairs

To achieve this, the hand must not contain the joker, since a pair and the joker would make the higher hand of three of a kind. The number of ways is therefore unchanged and remains as $123\,552$.

Card high

We will deal with this out of turn so that the more difficult calculation for one pair can be easily performed. The hand is only possible if it does not contain the joker and is naturally card high; we saw that this happens in 1 302 540 possible ways.

One pair

We can now calculate this by using the subtraction principle:

$$\binom{53}{5} - 13 - 204 - 3120 - 6552 - 7804 - 20\,532$$
$$- 137\,280 - 123\,552 - 1\,302\,540 = 1\,268\,088$$

ways.

The calculations result in table 10.3, where the frequencies are now divided by $\binom{53}{5} = 2\,869\,685$ to arrive at the probabilities.

Certainly, the odds have changed and, most significantly, for two pairs and three of a kind: these have been altered from 20 : 1 and 46 : 1 to 22 : 1 and 20 : 1 respectively. Three of a kind is now more likely than two pairs! If we are to rank the hands according to the likelihood of them being dealt we must then reverse their places in the table. But look at the effect that this would have on the player being dealt a hand which, with the joker, would count as three of a kind; it would be sensible for him to forgo the three-of-a-kind interpretation of the hand in favour of the two pairs alternative. This would mean that the only achievable three of a kind is the natural one, possible in 54 912 ways, whilst two pairs would now be achievable in 82 368 extra ways to add to the natural 123 552. The odds for two pairs and three of a kind then become 13 : 1 and 34 : 1 respectively, and again the order is reversed. We are left with an irreconcilable dilemma: based on frequency of occurrence, the hands cannot be properly ordered.

With two jokers the situation is even worse, as we can see from table 10.4. Once again three of a kind and two pairs are in the wrong order, but then so are one pair and a hand holding just an odd card. Added to this, four of a kind and a full house are

Table 10.3. Frequencies of poker hands with a single joker.

Hand	Frequency	Probability	Odds
Five of a kind	13	0.000 004 5	220 744 : 1
Straight flush	204	0.000 071	14 083 : 1
Four of a kind	3 120	0.001 087	919 : 1
Full house	6 552	0.002 283	437 : 1
Flush	7 804	0.002 72	367 : 1
Straight	20 532	0.007 15	139 : 1
Three of a kind	137 280	0.047 83	20 : 1
Two pairs	123 552	0.043 05	22 : 1
One pair	1 268 088	0.441 89	1.26 : 1
Odd card	1 302 540	0.453 90	1.20 : 1

Table 10.4. Frequencies of poker hands with two jokers.

Hand	Frequency	Probability	Odds
Five of a kind	78	0.000 025	39 999 : 1
Straight flush	564	0.000 179	5 586 : 1
Four of a kind	9 360	0.002 960	337 : 1
Full house	9 360	0.002 960	337 : 1
Flush	11 448	0.003 620	275 : 1
Straight	30 540	0.009 657	103 : 1
Three of a kind	233 584	0.073 860	12.5 : 1
Two pairs	123 552	0.039 068	24.6 : 1
One pair	1 440 464	0.455 481	1.2 : 1
Odd card	1 303 560	0.412 192	1.4 : 1

of equal precedence. As before, a player taking into account any of these facts will cause chaos in the ranking of the hands.

Of course, other options for wild cards are possible; 'deuces (twos) wild', for example; whatever the system in use, the problem remains that a player often has a choice of how to declare a hand, and that choice will invariably produce the strongest possible combination according to the accepted rules. Just how much wild cards alter the game has been analysed many times. An example in an article in *Chance* magazine (J. Emert and D. Umbach, 1996, Inconsistencies of 'wild-card' poker, *Chance*

9(3):17). The authors' analysis of wild-card poker variants concludes with the statement

> When wild cards are allowed, there is no ranking of the hands that can be formed for which more valuable hands occur less frequently.

For example, with *deuces wild*, they show that four of a kind occurs more than twice as often as a full house, yet modifying the rules as above leads to the same contradictory situation. That said, the authors examine several wild-card options and find that the standard ranking proves to have fewer inconsistencies than other possible ranking schemes.

Does an acceptable ranking scheme exist which is not marred by these problems? Emert and Umbach propose one based on what they define as 'the inclusion frequency' of a hand, which is a measure of the number of different winning combinations that can be declared from a given hand. Their method leads to the traditional rankings in poker without wild cards and does deal with the ambiguity introduced with the introduction of wild cards—but whether it will ever catch on with poker players is quite another matter!

Chapter 11

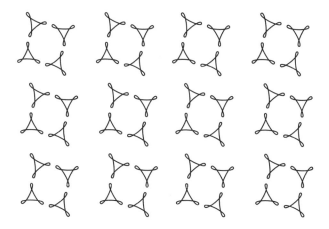

TWO SERIES

As the finite encloses an infinite series
And in the unlimited limits appear,
So the soul of immensity dwells in minutia
And in narrowest limits no limits inhere.
What joy to discern the minute in infinity!
The vast to perceive in the small, what divinity!

Jacob Bernoulli

Torricelli's Tower

In chapter 8 of *Nonplussed!* we considered the paradoxical solid
known (in particular) as *Torricelli's Trumpet*. Figure 11.1 shows
what this remarkable object looks like. It is formed as the solid of
revolution of the curve $y = 1/x$ for (say) $x \geqslant 1$ and calculus was
used to show that the trumpet had a finite volume but infinite

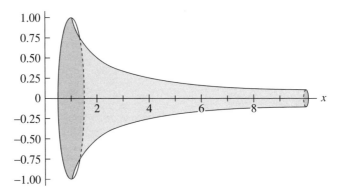

Figure 11.1.

surface area, the required computations being as follows: the volume is

$$\lim_{N\to\infty} \pi \int_1^N \left(\frac{1}{x}\right)^2 dx = \lim_{N\to\infty} \pi \int_1^N \frac{1}{x^2} dx$$

$$= \lim_{N\to\infty} \pi \left[-\frac{1}{x}\right]_1^N$$

$$= \lim_{N\to\infty} \pi \left(1 - \frac{1}{N}\right)$$

$$= \pi$$

and the surface area is

$$\lim_{N\to\infty} \int_1^N \frac{1}{x}\sqrt{1 + \frac{1}{x^4}}\, dx$$

$$= \lim_{N\to\infty} \int_1^N \frac{\sqrt{x^4 + 1}}{x^3}\, dx$$

$$= \lim_{N\to\infty} \left\{ -\frac{1}{2N^2}\sqrt{N^4 + 1} + \tfrac{1}{2}\ln(N^2 + \sqrt{N^4 + 1}) \right.$$

$$\left. + \frac{\sqrt{2}}{2} - \tfrac{1}{2}\ln(1 + \sqrt{2}) \right\},$$

which does not exist, since the second term is unbounded for large N.

The first calculation is easy and the second comparatively hard, requiring the techniques of integration by parts and substitution, and it is true that (as we observed) we could have made

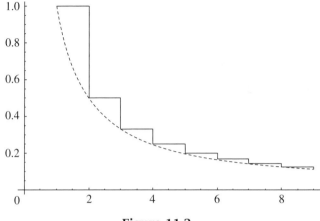

Figure 11.2.

life easier for ourselves in the second integral by noting that

$$\int_1^N \frac{1}{x}\sqrt{1 + \frac{1}{x^4}}\, dx > \int_1^N \frac{1}{x}\, dx = [\ln x]_1^N,$$

which assures the divergence.

In fact, we could have avoided calculus altogether if we wished to establish the nature (but not the exact value) of the volume and surface area by enclosing the trumpet in a horizontal tower, reminiscent of a tiered wedding cake laid on its side, generated by the piecewise function

$$f(x) = \begin{cases} 1, & 1 \leqslant x < 2, \\ \frac{1}{2}, & 2 \leqslant x < 3, \\ \frac{1}{3}, & 3 \leqslant x < 4, \\ \vdots & \vdots \\ \frac{1}{n}, & n \leqslant x < n+1, \end{cases}$$

rotated about the x-axis. Figure 11.2 shows this function (with the vertical line segments included) superimposed on a drawing of $y = 1/x$ and figure 11.3 shows the infinite tower generated by rotation about the x-axis.

Each horizontal line segment generates an element of surface area strictly less than the corresponding segment of the curve

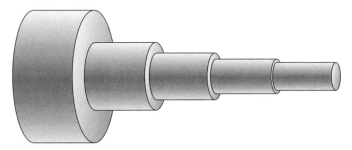

Figure 11.3.

and so calculating the surface area of the tower will be under-estimating the surface area of the trumpet. This computation is

$$\text{Surface area of trumpet} > \sum_{n=1}^{\infty} 2\pi \left(\frac{1}{n}\right) \times 1 = 2\pi \sum_{n=1}^{\infty} \frac{1}{n}.$$

If we wish to be precise and add in the areas of the annular tops of the sections, we simply add in the full area of the base of the first section—and so add in π.

To deal with the volume we realize that the trumpet is entirely contained in the tower and so the volume of the tower is a strict overestimate of the volume of the trumpet (which we know to be π anyway). Now the computation is

$$\text{Volume of trumpet} < \sum_{n=1}^{\infty} \pi \left(\frac{1}{n}\right)^2 \times 1 = \pi \sum_{n=1}^{\infty} \frac{1}{n^2}.$$

The estimates reduce the summation of two infinite series:

$$\sum_{n=1}^{\infty} \frac{1}{n} \quad \text{and} \quad \sum_{n=1}^{\infty} \frac{1}{n^2}.$$

The first sum is universally known as the *harmonic series*, since every term beyond the first is the harmonic mean of its two neighbours, where the harmonic mean of x and y is defined to be

$$\frac{2}{1/x + 1/y}.$$

Here we have

$$\frac{2}{\dfrac{1}{1/(n-1)}+\dfrac{1}{1/(n+1)}} = \frac{2}{n-1+n+1} = \frac{1}{n}.$$

The second sum has no standardized name but can reasonably be called *Euler's series*.

Since the advent of Zeno's *dichotomy paradox*, the idea of an infinite series converging to a finite sum has tested the mathematical philosopher—that its terms can become arbitrarily small and yet its sum can become arbitrarily large has tested the understanding of the infinite more deeply still. Zeno's simple argument that, before an object can travel a given distance d, it must travel a distance $d/2$ and to do this it must first travel a distance $\frac{1}{2}(d/2) = d/4$ can be continued indefinitely to the conclusion that the full distance can never be traversed. The apparent paradox was eventually resolved using the theory of infinite geometric series giving $\sum_{r=1}^{\infty} 1/2^r$ to be exactly 1. The infinite number of 'half-steps' needed is balanced by the progressively lesser amount of time needed to traverse the increasingly shorter distances and eventually the pure mathematical model of a physical situation breaks down. The two series mentioned above have their own place in the history of the infinite and the gallery of the surprising.

The Harmonic Series

Today I said to the calculus students, 'I know, you're looking at this series and you don't see what I'm warning you about. You look at it and you think, "I trust this series. I would take candy from this series. I would get in a car with this series." But I'm going to warn you, this series is out to get you. Always remember: The harmonic series diverges. Never forget it.'

So wrote the Manchester University mathematician Alexandre Borovik of the series $H_\infty = 1 + \frac{1}{2} + \frac{1}{3} + \frac{1}{4} + \frac{1}{5} + \cdots$. It is this series that is unveiled to almost every student beginning a course on real analysis as the canonical example of a series whose terms approach zero yet which diverges. It is also peculiarly difficult in that there is no usable, explicit formula for the exact sum to any given number of terms. The distinguished mathematician James Gregory, who for his time enjoyed a deep understanding of infinite series and their convergence, wrote in a letter dated 15 February 1671:

> As to yours, dated 24 Dec., I can hardly beleev, till I see it, that there is any general, compendious & geometrical method for adding an harmonical progression....

It is obvious that $H_1 = 1$, $H_2 = 1.5$ and $H_6 = 2.45$, but it takes progressively and noticeably more computer power to determine that $H_{100} = 5.187\ldots$, $H_{1000} = 7.486\ldots$ and $H_{1\,000\,000} = 14.392\ldots$.

Notice the three dots at the end of each of the last three sums; they are correct as far as they go but none can be written exactly since it can be proved that for all n other than 1, 2 and 6, H_n is always a nonterminating decimal. Above all, notice that the sums are small, a fact forcefully emphasized by the work of John W. Wrench Jr and Ralph. P. Boas Jr, who found the smallest n such that $H_n > 100$; that n is

15 092 688 622 113 788 323 693 563 264 538 101 449 859 497

(Partial sums of the harmonic series, 1971, *American Mathematical Monthly* 78:864–70). This glacially slow increase in the size of H_n strongly encourages the thought that it *must* converge, and to a pretty small number. That it does not was first established by the fourteenth-century French polymath Nicholas Oresme and subsequently by numerous others.

The Harmonic Series Diverges

We will give something of a flavour of the way in which the divergence has been approached over the centuries by reproducing four proofs, dating from the original to the modern day.

Oresme's fourteenth-century proof was as follows. Write the infinite sum with its terms collected as

$$H_\infty = 1 + \tfrac{1}{2} + (\tfrac{1}{3} + \tfrac{1}{4}) + (\tfrac{1}{5} + \tfrac{1}{6} + \tfrac{1}{7} + \tfrac{1}{8})$$
$$+ (\tfrac{1}{9} + \tfrac{1}{10} + \tfrac{1}{11} + \tfrac{1}{12} + \tfrac{1}{13} + \tfrac{1}{14} + \tfrac{1}{15} + \tfrac{1}{16}) + \cdots,$$

where the brackets consist of the 2^n terms which end with $1/2^{n+1}$ for $n = 1, 2, 3, \ldots$. This means that, taking the smallest term in each bracket,

$$H_\infty > 1 + \tfrac{1}{2} + (\tfrac{1}{4} + \tfrac{1}{4}) + (\tfrac{1}{8} + \tfrac{1}{8} + \tfrac{1}{8} + \tfrac{1}{8})$$
$$+ (\tfrac{1}{16} + \tfrac{1}{16} + \tfrac{1}{16} + \tfrac{1}{16} + \tfrac{1}{16} + \tfrac{1}{16} + \tfrac{1}{16} + \tfrac{1}{16}) + \cdots$$
$$= 1 + \tfrac{1}{2} + \tfrac{2}{4} + \tfrac{4}{8} + \tfrac{8}{16} + \cdots$$
$$= 1 + \tfrac{1}{2} + \tfrac{1}{2} + \tfrac{1}{2} + \tfrac{1}{2} + \cdots,$$

which is, of course, divergent.

Moving to the seventeenth century, Pietro Mengoli made implicit use of the harmonic relationship between consecutive terms, with the harmonic mean of a set of distinct numbers strictly less than their arithmetic mean. If the series is grouped as

$$H_\infty = 1 + (\tfrac{1}{2} + \tfrac{1}{3} + \tfrac{1}{4}) + (\tfrac{1}{5} + \tfrac{1}{6} + \tfrac{1}{7}) + (\tfrac{1}{8} + \tfrac{1}{9} + \tfrac{1}{10}) + \cdots,$$

then, since

$$\frac{1}{n-1} + \frac{1}{n+1} = \frac{2n}{n^2-1} > \frac{2n}{n^2} = \frac{2}{n},$$

the sum of the two outer terms in each triple is greater than twice the middle term, and this means that

$$H_\infty > 1 + \tfrac{3}{3} + \tfrac{3}{6} + \tfrac{3}{9} + \cdots = 1 + 1 + \tfrac{1}{2} + \tfrac{1}{3} + \cdots = 1 + H_\infty,$$

which yields an obvious contradiction.

In the eighteenth century, Jacob Bernoulli wrote the harmonic series, truncated by its first term, as the letter A. So,

$$A = \tfrac{1}{2} + \tfrac{1}{3} + \tfrac{1}{4} + \tfrac{1}{5} + \tfrac{1}{6} + \cdots$$

and, somewhat unnecessarily, the letter B as this series with numerators $1, 2, 3, 4, 5, \ldots$ So,

$$B = \tfrac{1}{2} + \tfrac{2}{6} + \tfrac{3}{12} + \tfrac{4}{20} + \tfrac{5}{30} + \cdots.$$

We will ignore this labelling and refer to both series as A.

Then he used the result of Leibniz that

$$1 + \tfrac{1}{3} + \tfrac{1}{6} + \tfrac{1}{10} + \tfrac{1}{15} + \cdots = 2,$$

which means that

$$\tfrac{1}{2} + \tfrac{1}{6} + \tfrac{1}{12} + \tfrac{1}{20} + \tfrac{1}{30} + \cdots = 1.$$

Subsequent letters were defined as follows:

$$C = \tfrac{1}{2} + \tfrac{1}{6} + \tfrac{1}{12} + \tfrac{1}{20} + \tfrac{1}{30} + \cdots = 1,$$
$$D = \tfrac{1}{6} + \tfrac{1}{12} + \tfrac{1}{20} + \tfrac{1}{30} + \cdots \qquad = C - \tfrac{1}{2} = 1 - \tfrac{1}{2} = \tfrac{1}{2},$$
$$E = \tfrac{1}{12} + \tfrac{1}{20} + \tfrac{1}{30} + \cdots \qquad = D - \tfrac{1}{6} = \tfrac{1}{2} - \tfrac{1}{6} = \tfrac{1}{3},$$
$$F = \tfrac{1}{20} + \tfrac{1}{30} + \cdots \qquad = E - \tfrac{1}{12} = \tfrac{1}{3} - \tfrac{1}{12} = \tfrac{1}{4},$$
$$G = \tfrac{1}{30} + \cdots \qquad = F - \tfrac{1}{20} = \tfrac{1}{4} - \tfrac{1}{20} = \tfrac{1}{5},$$
$$\vdots \qquad\qquad\qquad \vdots$$

Adding the left-hand column vertically, the middle column diagonally and the far right column vertically gives

$$C + D + E + F + G + \cdots$$
$$= \tfrac{1}{2} + (\tfrac{1}{6} + \tfrac{1}{6}) + (\tfrac{1}{12} + \tfrac{1}{12} + \tfrac{1}{12}) + (\tfrac{1}{20} + \tfrac{1}{20} + \tfrac{1}{20} + \tfrac{1}{20})$$
$$+ (\tfrac{1}{30} + \tfrac{1}{30} + \tfrac{1}{30} + \tfrac{1}{30} + \tfrac{1}{30}) + \cdots$$
$$= \tfrac{1}{2} + \tfrac{2}{6} + \tfrac{3}{12} + \tfrac{4}{20} + \tfrac{5}{30} + \cdots$$
$$= A = 1 + \tfrac{1}{2} + \tfrac{1}{3} + \tfrac{1}{4} + \tfrac{1}{5} + \cdots$$
$$= 1 + A.$$

Bernoulli concluded from this that 'the whole equals the part'. Although we saw in chapter 7 that this is possible, here he had an inescapable contradiction. This result appeared in the posthumously published *Ars Conjectandi,* together with the significant realization that 'the sum of an infinite series whose final term vanishes perhaps is infinite, perhaps finite', an uncomfortable realization that acted as muse for the verse at the head of this chapter.

As a final, modern demonstration, Ross Honsberger in his 1976 publication *Mathematical Gems II* (*Mathematical Association of America*) had the reader consider the following argument:

$$e^{H_n} = e^{1+1/2+1/3+1/4+1/5+\cdots+1/n}$$
$$= e^1 \times e^{1/2} \times e^{1/3} \times e^{1/4} \times \cdots \times e^{1/n}.$$

Since, for $x > 0$, $e^x > 1 + x$, it must be that

$$e^{H_n} > (1+1) \times \left(1 + \frac{1}{2}\right) \times \left(1 + \frac{1}{3}\right) \times \left(1 + \frac{1}{4}\right) \times \cdots \times \left(1 + \frac{1}{n}\right)$$
$$= \left(\frac{2}{1}\right) \times \left(\frac{3}{2}\right) \times \left(\frac{4}{3}\right) \times \cdots \times \left(\frac{n+1}{n}\right) = n + 1,$$

which means that e^{H_n} and therefore H_n are unbounded as n increases.

The fact that the harmonic series diverges has surprised and sometimes shocked many from the fourteenth to the twenty-first centuries, with repercussions extending far beyond Torricelli's trumpet, as the reader may wish to investigate.

Euler's Series

The volume of the trumpet involved our second series, which, curiously, has no generally accepted name. In this case it is easy to see that it does converge:

$$S_\infty = 1 + \frac{1}{2^2} + \frac{1}{3^2} + \frac{1}{4^2} + \cdots$$
$$= 1 + \left(\frac{1}{2^2} + \frac{1}{3^2}\right) + \left(\frac{1}{4^2} + \frac{1}{5^2} + \frac{1}{6^2} + \frac{1}{7^2}\right) + \cdots,$$

where the brackets consist of the 2^n terms that begin with $1/2^{2n}$.

This means that

$$S_\infty < 1 + \frac{2}{2^2} + \frac{4}{4^2} + \cdots$$

$$= 1 + \frac{1}{2} + \left(\frac{1}{2}\right)^2 + \left(\frac{1}{2}\right)^3 + \cdots = \frac{1}{1 - 1/2} = 2$$

with the final series an infinite geometric one of common ratio $\frac{1}{2}$. So, the series does converge, and to a number less than 2. The natural question to then ask is: what is the exact sum to infinity?

The problem of summing this series dates back to 1644 when Pietro Mengoli was once more involved when he asked what, precisely, that sum is. Subsequently, the problem was attacked by a veritable Who's Who of mathematicians, including John Wallis, Gottfried von Leibnitz and Jacob Bernoulli, who wrote in his 1689 publication *Tractatus de Seriebus Infinitis*, published in Basel,

> If anyone finds and communicates to us that which thus far
> has eluded our efforts, great will be our gratitude

and so the problem of identifying the exact sum of the infinite series has become known as the 'Basel Problem', 'the scourge of analysts', according to Montuela.

Once again, there is no explicit formula for the sum to n terms and the convergence is very slow indeed, which makes an accurate estimate of the sum difficult to find—and this makes the recognition of the number to which it is converging itself very difficult. John Wallis, the best British mathematician before Newton, had calculated its value as 1.645, which was extremely impressive since, simply by evaluating the sum, modern mathematical software demonstrates that with

$$S_n = 1 + \frac{1}{2^2} + \frac{1}{3^2} + \frac{1}{4^2} + \frac{1}{5^2} + \cdots + \frac{1}{n^2},$$
$$S_{100} = 1.634\,98\ldots,$$
$$S_{1000} = 1.643\,93\ldots,$$
$$S_{1\,000\,000} = 1.644\,93\ldots.$$

How easy the identification is for the modern researcher with such software and the use of N. J. A. Sloane's On-Line Encyclopedia of Integer Sequences (http://research.att.com/~njas/

sequences/). Typing in 1, 6, 4, 4, 9, 3 returns the only entry containing that sequence of digits: $\zeta(2) = \pi^2/6$.

It took the remarkable efforts of the Swiss genius Leonhard Euler to provide the astonishing answer, which he achieved in 1735, and so we will refer to the series using his name. The quite amazing fact that the sum involves π was as surprising to him as it is to the modern eye on first sight of it. Euler wrote, 'quite unexpectedly I have found an elegant formula involving the quadrature of the circle', by which he meant π.

Euler arrived at his solution only after Jacob Bernoulli's death, which, after having seen the resolution of the problem, brought the comment from his younger brother Johann (who had been mentor to the 28-year-old Euler): 'If only my brother were alive!'

Euler's Famous Proof

The Fundamental Theorem of Algebra has already appeared several times in these pages and, moving to the nineteenth century, Karl Weierstrass extended it to 'well-behaved' functions defined over complex numbers with the Weierstrass Factorization Theorem. In essence this tells us that we can, under certain conditions, 'factorize' such a function using its infinite number of zeros, as we can a polynomial with its finite number of them. In particular,

$$\sin \pi z = \pi z \prod_{n=1}^{\infty} \left(1 - \frac{z^2}{n^2} \right) \quad \text{for } z \in \mathbb{C}.$$

Euler foreshadowed the result with the typical nineteenth-century Weierstrassian mathematical rigour replaced by typical eighteenth-century Eulerian mathematical flamboyance. In fact, this was the third of the four proofs of the result that Euler produced and the first of elegance and significance.

We then have it that, if $\alpha_1, \alpha_2, \alpha_3, \ldots, \alpha_n$ are the roots of a polynomial $P_n(x)$ of degree n, then $x - \alpha_1, x - \alpha_2, x - \alpha_3, \ldots, x - \alpha_n$ are its factors, and so we have the identity

$$P_n(x) = A(x - \alpha_1)(x - \alpha_2)(x - \alpha_3) \cdots (x - \alpha_n).$$

Working in radians, Euler argued that the function $\sin x$ has the infinite set of roots $0, \pm\pi, \pm 2\pi, \pm 3\pi, \ldots$ and so, if we treat the function as a polynomial of 'infinite degree',

$$\begin{aligned}\sin x &= Ax(x - \pi)(x + \pi)(x - 2\pi)(x + 2\pi)(x - 3\pi)(x + 3\pi) \cdots \\ &= Ax(x^2 - \pi^2)(x^2 - 4\pi^2)(x^2 - 9\pi^2) \cdots.\end{aligned}$$

Now rewrite this as

$$Bx\left(1 - \frac{x^2}{\pi^2}\right)\left(1 - \frac{x^2}{2^2\pi^2}\right)\left(1 - \frac{x^2}{3^2\pi^2}\right) \cdots,$$

where B is a constant to be determined. With the angle measured in radians we have the result that

$$\frac{\sin x}{x} \to 1 \quad \text{as } x \to 0.$$

Dividing both sides by x and taking the limit across that infinite product of terms(!) allow us to evaluate B as 1, consequently,

$$\sin x = x\left(1 - \frac{x^2}{\pi^2}\right)\left(1 - \frac{x^2}{2^2\pi^2}\right)\left(1 - \frac{x^2}{3^2\pi^2}\right) \cdots.$$

Having developed this infinite product form of $\sin x$ he then used the infinite series form of the function, the Taylor expansion, which is valid for all x:

$$\sin x = x - \frac{x^3}{3!} + \frac{x^5}{5!} - \frac{x^7}{7!} + \cdots.$$

$\sin x$ has now done its job and he equated the series and the product to finish with

$$x - \frac{x^3}{3!} + \frac{x^5}{5!} - \frac{x^7}{7!} + \cdots = x\left(1 - \frac{x^2}{\pi^2}\right)\left(1 - \frac{x^2}{2^2\pi^2}\right)\left(1 - \frac{x^2}{3^2\pi^2}\right) \cdots.$$

We can easily see that the x terms on both sides are the same; it is the x^3 terms that are of real interest, with the series telling us that the coefficient is $-\frac{1}{3!}$ and the product (more subtly) telling us that it is the infinite series

$$-\frac{1}{\pi^2} - \frac{1}{2^2\pi^2} - \frac{1}{3^2\pi^2} - \frac{1}{4^2\pi^2} - \cdots;$$

the two must be the same and so

$$-\frac{1}{3!} = -\frac{1}{\pi^2} - \frac{1}{2^2\pi^2} - \frac{1}{3^2\pi^2} - \frac{1}{4^2\pi^2} - \cdots;$$

and tidying this up results in the final expression

$$\frac{1}{1^2} + \frac{1}{2^2} + \frac{1}{3^2} + \cdots = \frac{\pi^2}{6}.$$

How very remarkable.

Ten years after the proof, Euler wrote, 'The method was new and never used yet for such a purpose', but subsequently use it he did—many times over and some results gleaned from its use added weight to it.

For example, putting $x = \pi/2$ in his identity for $\sin x$ gives

$$\sin\frac{\pi}{2} = \frac{\pi}{2}\left(1 - \frac{1}{4}\right)\left(1 - \frac{1}{16}\right)\left(1 - \frac{1}{36}\right)\cdots$$

and so

$$1 = \frac{\pi}{2} \times \frac{3}{4} \times \frac{15}{16} \times \frac{35}{36} \times \cdots,$$

which can be rewritten as

$$\frac{2}{\pi} = \frac{1 \times 3 \times 3 \times 5 \times 5 \times 7 \times 7 \times \cdots}{2 \times 2 \times 4 \times 4 \times 6 \times 6 \times \cdots},$$

a result known to John Wallis a century earlier and here arrived at by this novel method quite unknown to Wallis.

The function $f(x) = 1 - \sin x$ provided another justification. It has its zeros at repeated multiples of $\pi/2$, as is demonstrated in figure 11.4.

Factorizing $1 - \sin x$ as before we then have

$$1 - \sin x = A\left(x - \frac{\pi}{2}\right)^2\left(x + \frac{3\pi}{2}\right)^2\left(x - \frac{5\pi}{2}\right)^2\left(x + \frac{7\pi}{2}\right)^2\cdots,$$

which can be transformed to

$$1 - \sin x = B\left(1 - \frac{2x}{\pi}\right)^2\left(1 + \frac{2x}{3\pi}\right)^2\left(1 - \frac{2x}{5\pi}\right)^2\left(1 + \frac{2x}{7\pi}\right)^2\cdots.$$

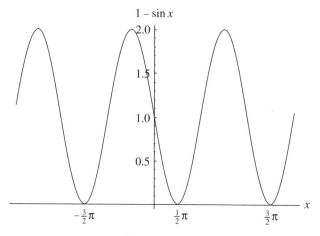

Figure 11.4.

And so, again using Taylor expansion,

$$1 - x + \frac{x^3}{3!} - \frac{x^5}{5!} + \frac{x^7}{7!} - \cdots$$
$$= B\left(1 - \frac{2x}{\pi}\right)^2 \left(1 + \frac{2x}{3\pi}\right)^2 \left(1 - \frac{2x}{5\pi}\right)^2 \left(1 + \frac{2x}{7\pi}\right)^2 \cdots ,$$

from which it is clear that B must be 1 and, comparing coefficients of x, gives

$$-1 = -\frac{4}{\pi} + \frac{4}{3\pi} - \frac{4}{5\pi} + \frac{4}{7\pi} - \cdots$$

and so

$$\frac{\pi}{4} = 1 - \frac{1}{3} + \frac{1}{5} - \frac{1}{7} + \frac{1}{9} - \cdots ,$$

a result which had already been established quite independently (and rigorously) by Leibniz. From this Euler commented:

> For our method, which may appear to some as not reliable enough, a great confirmation here comes to light.

Although there is mathematical alchemy here, and Euler well knew it, the signs were right that all was well—and eventually Weierstrass et al. would show it to be so.

A Criticism and a Rebuttal

Of the many arguments railed against Euler's novel but effective method, one of the most subtle was advanced by Daniel Bernoulli (Johann's son, friend and long-term correspondent of Euler), who suggested that $\sin x = 0$ may have complex roots and therefore the factorization may not be complete. This at a time, as we saw in chapter 5, when such things were far from understood. We cannot pass without looking at Euler's remarkable response, which provided the following (nearly complete) argument.

First, he needed his result that

$$\lim_{n \to \infty} \left(1 + \frac{x}{n}\right)^n = e^x$$

and second that

$$\sin x = \frac{1}{2i}(e^{ix} - e^{-ix}) \quad \text{and} \quad \cos x = \frac{1}{2}(e^{ix} + e^{-ix});$$

these are discussed in the appendix (page 225).

Now define

$$P_n(x) = \frac{1}{2i}\left[\left(1 + \frac{ix}{n}\right)^n - \left(1 - \frac{ix}{n}\right)^n\right],$$

then $\sin x = \lim_{n \to \infty} P_n(x)$.

He argued that this polynomial can have no complex roots since

$$P_n(x) = 0 \quad \Longleftrightarrow \quad \left(1 + \frac{ix}{n}\right)^n = \left(1 - \frac{ix}{n}\right)^n$$

$$\Longleftrightarrow \quad 1 + \frac{ix}{n} = e^{2k\pi i/n}\left(1 - \frac{ix}{n}\right).$$

This gives

$$x = \frac{n}{i}\frac{e^{k\pi i/n} - e^{-k\pi i/n}}{e^{k\pi i/n} + e^{-k\pi i/n}} = n\frac{(e^{k\pi i/n} - e^{-k\pi i/n})/2i}{(e^{k\pi i/n} + e^{-k\pi i/n})/2} = n\tan\frac{k\pi}{n},$$

which is real!

His extrapolation to the limit is again slightly shaky but it is a marvellous rebuttal.

A Rigorous Proof

To answer his critics in 1741 he published a fourth and more acceptable proof, which we give below. Certainly, it is more secure and in looking at it we can taste the nectar of mathematics produced by one of its master practitioners.

His method was to change the problem to one of summing over the odd terms of the series and then finding an expression for that sum. So,

$$
\begin{aligned}
S_\infty &= \frac{1}{1^2} + \frac{1}{2^2} + \frac{1}{3^2} + \frac{1}{4^2} + \frac{1}{5^2} + \frac{1}{6^2} + \cdots \\
&= \left(\frac{1}{1^2} + \frac{1}{3^2} + \frac{1}{5^2} + \cdots \right) + \left(\frac{1}{2^2} + \frac{1}{4^2} + \frac{1}{6^2} + \cdots \right) \\
&= \left(\frac{1}{1^2} + \frac{1}{3^2} + \frac{1}{5^2} + \cdots \right) + \frac{1}{4}\left(1 + \frac{1}{2^2} + \frac{1}{3^2} + \cdots \right) \\
&= \left(\frac{1}{1^2} + \frac{1}{3^2} + \frac{1}{5^2} + \cdots \right) + \frac{1}{4}S_\infty,
\end{aligned}
$$

which means that

$$
S_\infty = \frac{4}{3}\left(\frac{1}{1^2} + \frac{1}{3^2} + \frac{1}{5^2} + \cdots \right)
$$

and this changes the problem to finding an exact expression for

$$
\frac{1}{1^2} + \frac{1}{3^2} + \frac{1}{5^2} + \cdots .
$$

To do this, he considered the integral

$$
\int_0^1 \frac{\sin^{-1} t}{\sqrt{1 - t^2}}\, dt,
$$

which can be evaluated exactly to give

$$
\int_0^1 \frac{\sin^{-1} t}{\sqrt{1 - t^2}}\, dt = [\tfrac{1}{2}(\sin^{-1} t)^2]_0^1 = \tfrac{1}{2}(\sin^{-1} 1)^2 = \frac{\pi^2}{8}
$$

and also, by use of the Taylor Series for the inverse sine function,

$$
\sin^{-1} t = t + \frac{t^3}{6} + \frac{3t^5}{40} + \frac{5t^7}{112} + \frac{35t^9}{1152} + \cdots ,
$$

giving

$$\int_0^1 \frac{\sin^{-1} t}{\sqrt{1 - t^2}} \, dt$$

$$= \int_0^1 \frac{1}{\sqrt{1 - t^2}} \left(t + \frac{t^3}{6} + \frac{3t^5}{40} + \frac{5t^7}{112} + \frac{35t^9}{1152} + \cdots \right) dt$$

$$= \int_0^1 \frac{t}{\sqrt{1 - t^2}} \, dt + \frac{1}{6} \int_0^1 \frac{t^3}{\sqrt{1 - t^2}} \, dt + \frac{3}{40} \int_0^1 \frac{t^5}{\sqrt{1 - t^2}} \, dt$$

$$+ \frac{5}{112} \int_0^1 \frac{t^7}{\sqrt{1 - t^2}} \, dt + \frac{35}{1152} \int_0^1 \frac{t^9}{\sqrt{1 - t^2}} \, dt + \cdots$$

and there is an infinite series of similar integrals to perform. Writing the general integral as

$$I_n = \int_0^1 \frac{t^n}{\sqrt{1 - t^2}} \, dt, \quad n \in \{1, 3, 5, 7, \ldots\},$$

we can use integration by parts to find a recurrence relation

$$I_{n+2} = \int_0^1 \frac{t^{n+2}}{\sqrt{1 - t^2}} \, dt$$

$$= \int_0^1 t^{n+1} \{ t(1 - t^2)^{-1/2} \} \, dt$$

$$= [-t^{n+1} \sqrt{1 - t^2}]_0^1 + (n + 1) \int_0^1 t^n \sqrt{1 - t^2} \, dt$$

$$= (n + 1) \int_0^1 t^n \frac{1 - t^2}{\sqrt{1 - t^2}} \, dt$$

$$= (n + 1) \int_0^1 \frac{t^n}{\sqrt{1 - t^2}} \, dt - (n + 1) \int_0^1 \frac{t^{n+2}}{\sqrt{1 - t^2}} \, dt$$

$$= (n + 1) I_n - (n + 1) I_{n+2}$$

and so

$$I_{n+2} = \frac{n + 1}{n + 2} I_n, \quad n \geqslant 1.$$

We now need to evaluate

$$I_1 = \int_0^1 \frac{t}{\sqrt{1 - t^2}} \, dt = [-\sqrt{1 - t^2}]_0^1 = 1.$$

and use the recurrence relation repeatedly to find that

$$\frac{\pi^2}{8} = \int_0^1 \frac{\sin^{-1} t}{\sqrt{1-t^2}} \, dt$$

$$= I_1 + \frac{1}{6}I_3 + \frac{3}{40}I_5 + \frac{5}{112}I_7 + \frac{35}{1152}I_9 + \cdots$$

$$= 1 + \frac{1}{6} \times \frac{2}{3} + \frac{3}{40} \times \frac{4}{5} \times \frac{2}{3} + \frac{5}{112} \times \frac{6}{7} \times \frac{4}{5} \times \frac{2}{3}$$

$$+ \frac{35}{1152} \times \frac{8}{9} \times \frac{6}{7} \times \frac{4}{5} \times \frac{2}{3} \times \cdots$$

$$= 1 + \frac{1}{3^2} + \frac{1}{5^2} + \frac{1}{7^2} + \frac{1}{9^2} + \cdots .$$

The series of odd terms is summed and so we have

$$S_\infty = \frac{4}{3} \times \frac{\pi^2}{8} = \frac{\pi^2}{6}.$$

Now everyone is satisfied.

Chapter 12

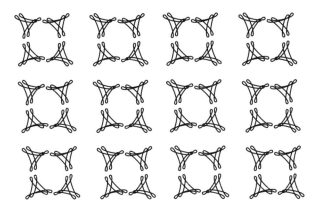

TWO CARD TRICKS

A mathematician is a conjurer who gives away his secrets.

John Conway

The bewilderment that accompanies a well-performed, good card trick relies greatly on the expertise of the conjurer—the ability to misdirect, to manipulate both cards and observer—and sometimes an underlying principle which is surprising in itself. We are interested in two such principles, both of which are the basis of numerous effects and both of which were discovered by academics.

The Kruskal Principle

The second chapter of Genesis continues the story of the creation of Heaven and Earth and begins (King James Version):

> Thus the heavens and the earth were finished and all the
> host of them And on the seventh day God ended his work
> which he had made and he rested on the seventh day from
> all his work which he had made And God blessed the sev-
> enth day and sanctified it because that in it he had rested
> from all his work which God created and made These are the
> generations of the heavens and **of** the earth when they were
> created in the day that the LORD God made the *earth* and
> the heavens And every plant of the field before it was in the
> earth and every herb of the field before it grew for the LORD
> God had not caused it to rain upon the earth and there was
> not a man to till the ground But there went up a mist from
> the earth and watered the whole face of the ground And
> the LORD **God** formed man of the dust of the ground and
> breathed into his nostrils the breath of life and man became
> a living soul...

and we can use the text to engage in a little 'Bible code' numer-
ology.

Start at the first word (*Thus*) and count along one word for each
of its four letters (to reach *the*); then count along one word for
each of its letters (to reach *finished*) and continue the process;
eventually the bold and underlined word **God** will be reached.
Coincidence perhaps? Now start at any word in (say) the first
six lines of the text and the same will happen: the sequence will
reach the same occurrence of the word God. In fact, if we start
at any word up to and including the **of** the same thing will happen. That all paths lead to God is comforting, but is this more
than coincidence? Actually, a nonmystical explanation is perfectly simple: the chains of words generated in the above manner
all have an instance of the word *earth* as a first common link, and
of course from that point on the chains are identical and since
one of them has *God* as a link, all must do so. The passage is
repeated below with that first common link and all subsequent
links emboldened:

> Thus the heavens and the earth were finished and all the
> host of them And on the seventh day God ended his work
> which he had made and he rested on the seventh day from

all his work which he had made And God blessed the seventh day and sanctified it because that in it he had rested
from all his work which God created and made These are
the generations of the heavens and <u>of</u> the earth when they
were created in the day that the LORD God made the **earth**
and the heavens And **every** plant of the field **before** it was
in the earth **and** every herb **of** the **field** before it grew for
the LORD God **had** not caused **it** to **rain** upon the earth **and**
there was **not** a man **to** till **the** ground But **there** went up
a mist **from** the earth and **watered** the whole face of the
ground **And** the LORD <u>God</u> formed man **of** the **dust** of the
ground **and** breathed into **his** nostrils the **breath** of life and
man became **a living** soul...

This matter was discussed in the August 1998 issue of *Scientific
American* but the underlying principle dates back further—and
was discussed in quite another context.

An article entitled 'Sum total' appeared in the December 1957
issue of the magic periodical *Ibidem*, written by the magician
Alexander F. Kraus, and which brought to the world of card
magic the 'Kraus Principle'.

The magician *M* asks a member of the audience *A* to shuffle
a standard pack of 52 playing cards, then secretly to pick an
integer between 1 and 10. *A* is then asked to deal the cards, one
by one and face up, to form a pile and while doing so to count
them silently in the following way.

Suppose that the chosen number is 6. The sixth card dealt
becomes a 'key' card, and its face value dictates how many more
cards must be dealt to arrive at the next key card. For example,
if the first key card happens to be 3, that many cards are dealt to
arrive at the next key card. The procedure is repeated, generating
a chain of key cards, until the pack is exhausted, and when this
happens it is likely that the count dictated by the final key card
will prove impossible to complete; this final key card is the secret
chosen card.

Of course, the magician's task is to identify that secret card
chosen by this random process, which seems a very big ask.

In the 1970s the Kraus Principle became the 'Kruskal Principle', when the late Princeton physicist Martin Kruskal rediscovered the novelty that the magician has a very good chance of
amazing the audience by announcing that secret card. The fact
is that any two such chains will, in all likelihood, come together
at some point, which means of course that they remain together
from that point on—and inevitably end with the same card. The
second chain is generated by the magician.

If we refer back to the 'Bible code' example, the passage may be
thought of as a set of cards with each card numbered according
to the number of letters in each word. Each starting word generates its own chain—and each chain meets at the 'card' named
'earth'.

In order to make headway with an analysis we will generalize to
a set of n cards which are numbered $\{c_1, c_2, c_3, \ldots, c_n\}$, and this
is, of course, the list of possible step sizes. Further, we will make
the simplifying assumption that the two chains are equally likely
to intersect at any of the n cards, with a constant probability
which, for later convenience, we will write as p^2. If we write $q =
1 - p^2$ and $I \in \{1, 2, 3, \ldots, n\}$ as the random variable which is
the card position at which they intersect, we have the standard
geometric probability distribution

$$P(I = r) = q^{r-1}p^2 \quad \text{for } r = 1, 2, 3, \ldots.$$

If we invoke a standard result of a geometric random variable
(which is easy to prove), we have that

$$P(I > r) = q^r.$$

This means that the probability that the magician succeeds with
the trick is

$$P = 1 - P(I > n) = 1 - q^n = 1 - (1 - p^2)^n.$$

Now we have to find a reasonable estimate for p^2 and to do this
we find such for p. For each of the two chains, the average step
size is $(1/n) \sum_{r=1}^{n} c_r$, which naturally generates a probability of
$1/((1/n) \sum_{r=1}^{n} c_r)$, and we will call p this probability. That is,

we will estimate the probability that the two chains meet on any card using the assumption that they are 'on average' realizing their average step lengths and so 'on average' will meet in such a manner. Actually, the technical analysis belongs to the world of Markov chains and the interested reader might wish to consult the October 2001 paper of J. Lagarias, E. Rains and R. Vanderbei, The Kruskal count.[1]

The Lagarias et al. argument is firmer and we produce a variant of part of it for later convenience.

Recall from chapter 6 the definition of conditional probability and the dissection of an event into its component parts to write

$$P(A \mid B) = \frac{P(A \cap B)}{P(B)}$$

and

$$P(E) = P(E \mid A)P(A) + P(E \mid B)P(B)$$
$$+ P(E \mid C)P(C) + P(E \mid D)P(D).$$

Then, for $r > s$, we have

$$P(I > r \mid I \geqslant s) = \frac{P(I > r \cap I \geqslant s)}{P(I \geqslant s)} = \frac{P(I > r)}{P(I \geqslant s)} = \frac{(1 - p^2)^r}{(1 - p^2)^{s-1}}$$
$$= (1 - p^2)^{r-s+1} = P(I > r - s + 1).$$

With all of this in place we examine the event that the two sequences have not intersected before the nth card, that is, we will find an expression for $P(I > n)$ (the event E above). To do this, define the random variables M_1 and A_1 to be the position of the first card chosen by the magician and the member of the audience respectively and divide all possibilities into the four categories (the events A, B, C and D) above:

$$A = (M_1 \geqslant 2 \cap A_1 \geqslant 2), \qquad B = (M_1 = 1 \cap A_1 \geqslant 2),$$
$$C = (M_1 \geqslant 2 \cap A_1 = 1), \qquad D = (M_1 = 1 \cap A_1 = 1).$$

---

[1]http://front.math.ucdavis.edu/search?a=lagarias&t=&q=&c=&n=40&s= Listings.

This results in the decomposition

$$P(I > n)$$
$$= [P(I > n \mid (M_1 \geqslant 2 \cap A_1 \geqslant 2))] \times P(M_1 \geqslant 2 \cap A_1 \geqslant 2)$$
$$+ [P(I > n \mid (M_1 = 1 \cap A_1 \geqslant 2))] \times P(M_1 = 1 \cap A_1 \geqslant 2)$$
$$+ [P(I > n \mid (M_1 \geqslant 2 \cap A_1 = 1))] \times P(M_1 \geqslant 2 \cap A_1 = 1)$$
$$+ [P(I > n \mid (M_1 = 1 \cap A_1 = 1))] \times P(M_1 = 1 \cap A_1 = 1).$$

The event $A = (M_1 \geqslant 2 \cap A_1 \geqslant 2)$ is precisely the event $I \geqslant 2$ and so, using the above result,

$$P(I > n \mid (M_1 \geqslant 2 \cap A_1 \geqslant 2)) = P(I > n \mid I \geqslant 2)$$
$$= P(I > n - 2 + 1)$$
$$= P(I > n - 1).$$

The same argument holds for each of the next two terms and the last term is 0 since

$$P(I > n \mid (M_1 = 1 \cap A_1 = 1)) = P(I > n \mid I = 1) = 0.$$

The expression therefore reduces to

$$P(I > n) = P(I > n - 1)[P(M_1 \geqslant 2 \cap A_1 \geqslant 2)$$
$$+ P(M_1 = 1 \cap A_1 \geqslant 2)$$
$$+ P(M_1 \geqslant 2 \cap A_1 = 1)]$$

and since the bracketed expression is simply $P(I > 1)$ we are left with the recurrence relation

$$P(I > n) = P(I > n - 1)[P(I > 1)] = P(I > n - 1)[1 - p^2]$$

and chasing this down results in

$$P(I > n) = P(I > n - 1)[P(I > 1)]$$
$$= P(I > 1)[1 - p^2]^{n-1}$$
$$= (1 - p^2)^n,$$

which means that the probability P of the magician performing the trick successfully is given by $P = 1 - (1 - p^2)^n$.

Now we can calculate these theoretical probabilities, making $n = 52$ and giving a value to p, which is, of course, entirely dependent on what values we ascribe to the cards. We will look at five reasonable alternatives.

- The spot cards are given their natural values and the jack, queen and king are given values 11, 12 and 13 respectively. The card values will be among $\{1, 2, 3, \ldots, 13\}$ and the average is 7. We take $p = \frac{1}{7}$ to get

$$P = 1 - P(I > 52) = 1 - [1 - (\tfrac{1}{7})^2]^{52}$$
$$= 1 - (\tfrac{48}{49})^{52} = 0.6577\ldots,$$

  which means that the chances of the chains meeting is about 66%.

- The spot cards are given their natural values and the court cards each count as 10. The card values will be among $\{1, 2, 3, \ldots, 10, 10, 10, 10\}$ and the average is $\frac{85}{13}$. We take $p = \frac{13}{85}$ to get

$$P = 1 - P(I > 52) = 1 - [1 - (\tfrac{13}{85})^2]^{52}$$
$$= 1 - (\tfrac{7056}{7225})^{52} = 0.7080\ldots,$$

  which means that the chances of the chains meeting is about 71%.

- The spot cards are given their natural values and the court cards each count as 5. The values will be among $\{1, 2, 3, \ldots, 9, 10, 5, 5, 5\}$ and the average is $\frac{70}{13}$. We take $p = \frac{13}{70}$ to get

$$P = 1 - P(I > 52) = 1 - [1 - (\tfrac{13}{70})^2]^{52}$$
$$= 1 - (\tfrac{4731}{4900})^{52} = 0.8388\ldots,$$

  which means that the chances of the chains meeting is now about 84%.

- The spot cards are given their natural values and the court cards each count as 1. The values will be among

$\{1, 2, 3, \ldots, 10, 1, 1, 1\}$ and the average is $\frac{58}{13}$. We take $p = \frac{13}{58}$ to get

$$P = 1 - P(I > 52) = 1 - [1 - (\tfrac{13}{58})^2]^{52}$$
$$= 1 - (\tfrac{3195}{3364})^{52} = 0.9315\ldots$$

and the chances of the chains meeting becomes a very impressive 93%.

As the average value of the cards decreases, so (very reasonably) the chances of two chains meeting increases and with this in mind we can consider a particularly deceptive card assignment.

- Ignore the numbers on the cards and use the number of letters in the card names for the card numbers (ace = 3, two = 3, $\ldots$, queen = 5, king = 4). Now the values will be among $\{3, 3, 5, 4, 4, 3, 5, 5, 4, 3, 4, 5, 4\}$ and the average is $\frac{52}{13} = 4$. We take $p = \frac{1}{4}$ in our estimate to get

$$P = 1 - P(I > 52) = 1 - [1 - (\tfrac{1}{4})^2]^{52}$$
$$= 1 - (\tfrac{15}{16})^{52} = 0.9651\ldots$$

and the chances of the chains meeting becomes an even more impressive 97%.

A little extra edge can be gained by the magician if he selects the top card of the pack, and we can use the Lagarias et al. argument to establish the details:

$P(I > n)$
$$= [P(I > n \mid (M_1 = 1 \cap A_1 \geqslant 2))] \times P(M_1 = 1 \cap A_1 \geqslant 2)$$
$$= P(I > n - 1)[P(I > 1)]$$

as before. Chasing this down results in

$$P(I > n) = P(I > 1)[1 - p^2]^{n-1}$$

but now $P(I > 1) = 1 - p$ since only one of the sequences has an arbitrary start and this means that $P(I > n) = (1 - p)[1 - p^2]^{n-1}$ and

$$P = 1 - (1 - p)[1 - p^2]^{n-1}.$$

Table 12.1. Theoretical probabilities.

Picture cards count as	Value of p	Magician starts with any of the first 10 cards	Magician starts with first card
11, 12, 13	$\frac{1}{7}$	65.77	70.05
10, 10, 10	$\frac{13}{85}$	70.80	74.67
5, 5, 5	$\frac{13}{70}$	83.88	86.41
1, 1, 1	$\frac{13}{58}$	93.15	94.40
Letter count	$\frac{1}{4}$	96.51	97.21

Table 12.2. Empirical probabilities.

Picture cards count as	Value of p	Magician starts with any of the first 10 cards	Magician starts with first card
11, 12, 13	$\frac{1}{7}$	67.93	69.46
10, 10, 10	$\frac{13}{85}$	70.27	72.46
5, 5, 5	$\frac{13}{70}$	84.29	85.35
1, 1, 1	$\frac{13}{58}$	93.66	94.30
Letter count	$\frac{1}{4}$	95.23	95.84

The same calculations as above give rise to the final column of table 12.1, which gives the chances of the chains meeting under the various card enumerations, given as a percentage.

With this mixture of exactitude and heuristics it can only be right to test the model with a computer simulation and table 12.2 shows the results of such a simulation over 100 000 trials in each case.

As we have said, more sophisticated mathematical techniques can be utilized to make more firm the heuristic parts of the argument but they are rather specialized and, without considerable preparation, rather opaque. We have already mentioned the CBS series *NUMB3RS* in chapters 6 and 8; the 16 February

2007 episode 'Contenders' had Amita and Charlie use (a modified form of) the Kruskal count to help hunt down the killer of
two boxers. Armed with this material, the reader might wish to
revisit that DVD with renewed insight!

The Gilbreath Principle. The magician sits at a table opposite
a spectator. He produces a pack of cards and casually spreads
them face up to prove that they are properly mixed; he gath-
ers them up, places them face down in a pack, cuts the pack
and is just about to riffle shuffle the two parts together when
he declares that the spectator should do this instead and then
cut the pack and complete the cut a few times. However well
or badly, the cards are riffle shuffled together, cut and returned
to the magician. He conceals the pack behind his back and pro-
ceeds to bring cards four at a time to the front and place them
in face-down piles on the table. When all thirteen piles of four
cards have been formed each pile is turned over, and in each
pile there is precisely one spade, one heart, one diamond and
one club.

The spectator shuffled and cut the cards, so how could the
magician tell which suit each is, even if he had arranged them
in some order not obvious as they were spread? The answer is
that the cards were indeed arranged and that the magician does
not know the suit of each card, but he does know how to utilize
Gilbreath's Principle.

In February 1957 an American magician with the name John
Russell Duck produced the first edition of his magic magazine
The Cardiste in which he included the article entitled 'The Rus-
duck stay-stack system'. It was the observation that, with any
number of *perfect* riffle shuffles of the top and bottom halves
of the pack (one in which cards from each of the two halves are
alternately placed one upon the other) some of the original order
is preserved. For example, if the cards were originally ordered as
ace to king in spades, hearts, clubs and diamonds, the following
would still be the case:

- each half of the pack will contain two cards of the same
denomination;

- each half of the pack will contain thirteen red cards and thirteen black cards;
- each half of the pack will contain one red-suited and one black-suited card of all thirteen denominations;
- the top card will be the same denomination as the bottom card, the second from top card will be the same denomination as the second from bottom card and so on throughout the pack.

We can see that a sequence of perfect riffle shuffles does preserve a great deal of order, and magicians over the years have put this fact to bewildering use. In fact, if we distinguish between the perfect out-shuffle (in which the top card stays on top of the pack) and the perfect in-shuffle (in which the top card moves to the second position down), just 8 perfect out-shuffles or 52 perfect in-shuffles will return the pack to its original order.

What if the riffle shuffles are not perfect? Bayer and Diaconis (D. Bayer and P. Diaconis, 1992, Trailing the dovetail shuffle to its lair, *Annals of Applied Probability* 2:294–313) proved the surprising result that it takes about eight riffle shuffles to make every configuration of the cards (approximately) equally likely and so randomize the pack (compare this number with the approximately 2500 overhand shuffles that are needed to randomize to the same extent); if we think of the $52! = 8.07 \times 10^{67}$ possible orderings of the cards, these small numbers are themselves surprising. Riffle shuffling more than eight times does not significantly increase the randomness, and riffle shuffling less than eight times is insufficient to ensure randomness. This last point is elegantly demonstrated by the following argument that five riffle shuffles are insufficient to randomize a pack of cards:

- We take the definition of *random* to be that every configuration is (approximately) equally likely and show that a particular configuration is not reachable by the process.
- Number the original cards 1 to 52, starting at the top.
- At any stage of the shuffling, a *rising sequence* in the pack is defined to be a (possibly broken) increasing sequence of consecutive integers of maximal length, so the original pack has just one rising sequence $1, 2, 3, \ldots, 51, 52$.

- As the original pack is randomly cut into two parts, two rising sequences are created: the top pack and the bottom pack. Riffle shuffling the two packs together simply intersperses one rising sequence with the other; as a demonstration, consider a pack of just eight cards $\{1, 2, 3, 4, 5, 6, 7, 8\}$ cut as $\{1, 2, 3, 4, 5\}$ and $\{6, 7, 8\}$ and riffle shuffle together to form the pack $\{\underline{1}, \overline{6}, \overline{7}, \underline{2}, \underline{3}, \underline{4}, \overline{8}, \underline{5}\}$.

- After the second riffle shuffle the number of rising sequences is as most $2 \times 2 = 4$, since each of the two rising sequences from the first shuffle has a chance of being cut into two. For example, $\{1, 6, 7, 2, 3, 4, 8, 5\}$ might be cut as $\{1, 6, 7, 2\}$ and $\{3, 4, 8, 5\}$ and shuffled together as $\{1, 3, 4, 8, 6, 7, 5, 2\}$ to result in the four rising sequences: $\{1, 2\}$, $\{3, 4, 5\}$, $\{8\}$, and $\{6, 7\}$.

- The same arguments holds for subsequent riffle shuffles and we conclude that a riffle shuffle at most doubles the number of rising sequences; after five riffle shuffles there will then be at most 32 rising sequences.

- Now consider the pack reversed, with card 52 at the top and card 1 at the bottom; this has 52 rising sequences, each of 1 card, and cannot possibly be reached in the five riffle shuffles.

It was the amateur magician Norman L. Gilbreath who deliberated on what structure remains after a single, imperfect riffle shuffle. In the June 1966 edition of another magicians' publication, *Linking Rings*, there appeared descriptions by him of what have become known as 'Gilbreath's first and second principles', which identify something which can be salvaged from the order of a pack of cards if the riffle shuffle is not perfect. In (nearly) his own words, his stated first principle was:

> If a deck of cards, ordered in alternating colours, is cut into two parts with the bottom cards of the two parts having opposite colours, and the two parts are riffle shuffled together then each successive pair of cards is composed of one red card and one black card.

And then:

> If two groups of similar cards, one in the reverse order of
> the other, are riffle shuffled together the two halves of the
> resulting group of cards are similar to the original group.

First, let us look at this second observation. By it he meant that,
if we take a set of n different cards, ordered in some way, and
then a second such pack but arranged in the opposite order,
riffle shuffle the two of them together and separate the pack
into the top and bottom equal piles, each pile will contain those
n different cards in some order; there will be no omissions or
repeats.

The reader might wish to experiment with this by, for example,
taking the two packs of 13 cards ace to king of spades and then
king to ace of hearts and riffling them together. Each half will
then contain both hearts and spades, but also the full set of ace
to king. Notice that the reversal of order can be accomplished
subtly by having the two half packs arranged in the same order,
one on top of the other, and then counting through the top half
pack one card at a time, placing the cards face down on top of
each other.

The analysis of this is not too hard. If we call the first pack
of cards, from top to bottom, $\{X_1, X_2, \ldots, X_n\}$ and the second
$\{X_n, X_{n-1}, \ldots, X_1\}$, they will combine to form a new pack of $2n$
cards which we then divide into the top half A and the bottom
half B. Now suppose that X_k is the last element from the reversed
pack $\{X_n, X_{n-1}, \ldots, X_1\}$ to appear in A, then so must all of the
set $\{X_n, X_{n-1}, \ldots, X_k\}$; this is $n - k + 1$ elements. It must be that
$\{X_1, X_2, \ldots, X_n\}$ contributes $n - (n - k + 1) = k - 1$ elements to
A, and these must be $\{X_1, X_2, \ldots, X_{k-1}\}$. This means that A must
consist of the distinct elements $\{X_1, X_2, \ldots, X_n\}$ and, of course,
so must B. It is clear that the whole matter depends on the fact
that a riffle shuffle may well mix cards up but it does preserve
relative order.

This analysis does not cope with the first principle, where there
is a repeated pattern of red and black cards, not two blocks of
identical cards in reverse order. In fact, Gilbreath's Principle (and
in reality there is only one) is precisely defined by the following
statement.

Suppose that $S = \{X_1, X_2, X_3, \ldots, X_n\}$ is a set of n cards which are, in some agreed way, different from each other. Now take multiple copies of S and form a pack by stacking them one upon the other. Form a second pack by taking (not necessarily the same number of) multiple copies, but arranging each copy of S in reverse order. Now there are two face-down packs, not necessarily of the same size; one consisting of repeats of $\{X_1, X_2, X_3, \ldots, X_n\}$ and the other of $\{X_n, X_{n-1}, X_{n-2}, \ldots, X_1\}$. Riffle shuffle the two packs together and divide the combined pack back into the sets of n cards, counting from the top. Within each set of these n cards, $\{X_1, X_2, X_3, \ldots, X_n\}$ will again appear in some order.

For example, if $n = 2$, $\{X_1, X_2\}$ might be the pairing {red card, black card}, which accounts for Gilbreath's first statement. With $n = 4$, $\{X_1, X_2, X_3, X_4\}$ might be the set $\{\clubsuit, \heartsuit, \spadesuit, \diamondsuit\}$, and the secret behind the magician's trick described earlier is revealed.

In fact, in the August 1966 issue of *Linking Rings*, one Charles Hudson really said just that:

> When a repeating series of cards is riffle shuffled into itself, with one of the packets to be shuffled being in reverse order to the other, the contents of each group in the series do not change—they are only disordered.

Again, the reader might wish to experiment with these black and red combinations or with the arrangement of the four suits; it does work, and again it is not too hard to see why. Whenever a last element X_k of a reversed pack is inserted into one of the non-reversed packs, then so is all of $\{X_n, X_{n-1}, \ldots, X_k\}$ and exactly that set of cards forms the final $n - k + 1$ cards of the forward pack and these are forced to the top of the next forward pack of n cards—and so the cascade continues. All that remains is to explain why, in Gilbreath's first statement about the red and black effect, the bottom pack is not reversed. In fact it is, by the statement 'is cut into two parts with the bottom card of the two parts having opposite colours'. Of course, this means that all corresponding pairs have opposite colours and that is just a special case of a reversal! The cutting of the pack has no real effect.

Finally, if the pack is arranged in any suitable multiply structured manner, the magician can wreak even greater havoc with the spectator's thoughts. For example, the repeated {♣, ♥, ♠, ♦} arrangement automatically allows for the magician to intersperse the production of red–black pairs with quartets of cards of each suit. Additionally, if a particular selection of denominations is chosen for the first thirteen cards and repeated three times, each consecutive set of thirteen cards will contain one card of each denomination: such an arrangement is shown below:

A♣, 8♥, 5♠, 4♦, J♣, 2♥, 9♠, 3♦, 7♣, Q♥, K♠, 6♦, T♣,
A♥, 8♠, 5♦, 4♣, J♥, 2♠, 9♦, 3♣, 7♥, Q♠, K♦, 6♣, T♥,
A♠, 8♦, 5♣, 4♥, J♠, 2♦, 9♣, 3♥, 7♠, Q♦, K♣, 6♥, T♠,
A♦, 8♣, 5♥, 4♠, J♦, 2♣, 9♥, 3♠, 7♦, Q♣, K♥, 6♠, T♦.

With the cards so arranged, top to bottom, a multiple effect would be:

- Have the spectator cut the cards and complete the cut at will.
- Ask for the cards to be dealt face down on top of each other until there is a decent pack on the table.
- Ask for the two packs to be riffle shuffled together.
- The magician produces pairs of red–black cards and quartets of different suits for the first twenty-six cards, after which the top remaining thirteen cards are dealt face up to reveal one card of each denomination, which is repeated with the final thirteen cards.

It's all a matter of where the magician places the emphasis.

Chapter 13

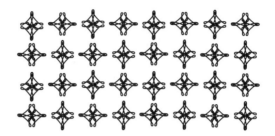

THE SPIN OF A NEEDLE

Geometry is the science of correct reasoning on incorrect figures.

George Pólya

In *Nonplussed!* we considered some consequences of tossing a needle, in particular, onto a set of equally spaced parallel lines. The fact, surprising and historically significant, that the probability of the needle landing across a line involved π was first investigated by the eighteenth-century French scientist Georges Louis Leclerc, Comte de Buffon: hence the name *Buffon's Needle.* Here we move to the twentieth century to consider a simple question about spinning the needle, which has its own intriguing and important answer.

A Japanese Question from Japan

In 1922–23 the eminent American mathematician George Birkoff (and father of Garrett) gave a series of lectures on the theory of relativity at the Lowell Institute and at the 'southern branch' of the University of California (now the University of California at Los Angeles). Their unsurprising success led to him 'revising, extending, and unifying the material in book form', which brought about the 1925 volume, *The Origin, Nature and Influence of Relativity*, the first chapter of which ('Euclid, Newton, Faraday and Einstein') contains the following:

> It may not be amiss to note in passing that not all elementary geometric conundrums have yet been answered by professional mathematicians. Thus, map-makers have noted that apparently any imaginable map on the plane or sphere can be coloured in only four colours in such wise that two countries with a common boundary line have different colours. Despite persistent efforts, the truth of this conjecture has not yet been established, although five colours are known to be enough. Of like intriguing simplicity is the question raised a few years ago by the Japanese mathematician Kakeya as to the least area within which a line of given length can be turned around in a plane. An area only half as great as that of the circle with this length for diameter will suffice. No one has as yet been able to prove that this is the least possible area.

It took until 1976 for K. Appel and W. Haken to provide their controversial, computer-assisted proof of the four-colour theorem: Birkoff, one of the foremost mathematicians of his time, could not have known it, but the second question had in effect been answered almost as soon as it had been asked, with that answer shrouded behind an impenetrable political veil.

In 1917 the Japanese mathematician Soichi Kakeya had asked the question:

> In the class of all figures in which a segment of length 1 unit can be turned through 180°, remaining always within the figure, which one has the smallest area?

Also in 1917 the Russian mathematician Abram Besicovitch had solved a seemingly different problem of his own. It would take several years before Besicovitch became aware of Kakeya's problem of 'intriguing simplicity' and so provide a resolution which is entirely remarkable and was entirely unexpected.

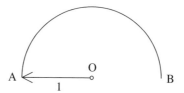

Figure 13.1.

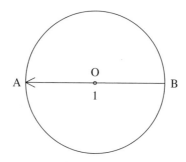

Figure 13.2.

Elementary Evidence

Our '*needle*' will be 'a segment of length 1 unit'.

The 180° rotation is easily achieved. Take a semicircular disc of diameter 2 units with the needle OA as a radius, as shown in figure 13.1.

Rotate OA by 180° about the centre O so that the arrowhead coincides with B and then translate the line horizontally to the left by 1 unit; the reversal has been accomplished in an area of $\frac{1}{2}(\pi \times 1^2) = \frac{1}{2}\pi = 1.5707\ldots$ (Note that the translation in the line of the needle has not taken up area.)

Yet the rotation can be achieved in half the area if we consider the needle as the diameter of a circular disc, as in figure 13.2.

Rotating the needle about its centre (the centre of the circle) achieves the desired result without translation, in an area of $\pi \times (\frac{1}{2})^2 = \frac{1}{4}\pi = 0.7853\ldots$

We can improve matters still further using an equilateral triangle of height 1 unit, as shown in figure 13.3.

Place the needle AB along the side XZ, with A at X. Rotate AB 60° anticlockwise about A/X so that it lies along the side XY, then translate AB along its direction so that B is at Y. Now

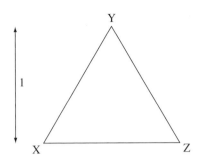

Figure 13.3.

repeat the rotation, this time about B/Y, and translate so that A
is at Z. One final rotation about A/Z and a translation so that
B coincides with X will have the needle to its original position
and completely turned it around. Has this extra complication
paid dividends? The answer is yes, since the side of the trian-
gle is of length $2/\tan 60° = \frac{2}{\sqrt{3}}$ and so the area of the triangle is
evidently $\frac{1}{2} \times \frac{2}{\sqrt{3}} \times 1 = \frac{1}{\sqrt{3}} = 0.5773\ldots$ (once again note that the
translations along the lines of the needle have not taken up any
area).

It is with this equilateral triangle that a part of the story ends.

A Hungarian Solution from Denmark

In 1919 Julius Pál, a highly gifted and ambitious Hungarian
Jew, finally moved from Pozsony in his native Hungary to
Copenhagen in Denmark, which was free from redolent political intrigue and in which he could find the position he sought
to enable him to carry out his research. Harold Bohr (brother of
Neils) was an influential sponsor and a link between Pál and that
other major character in this story, Abram Besicovitch.

Kakeya, his collaborator M. Fujiwara and others had immediately conjectured that the equilateral triangle was the *convex*
shape of minimal area to achieve the required purpose and Fujiwara or possibly Bohr communicated the problem to Pál, who
published a proof of the conjecture in 1921 (Ein Minimumprob-
lem für Ovale, 1921, *Mathematische Annalen* 83:311–19) which
put the problem to rest, but only for convex regions. The non-
convex version of the problem still remained open, and it was

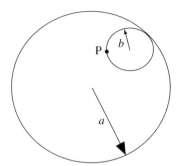

Figure 13.4.

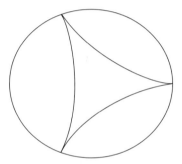

Figure 13.5.

this to which Birkoff referred in his book with his comment: 'An area only half as great as that of the circle with this length for diameter will suffice', referring rather obliquely to the deltoid, which is a particular member of the family of hypocycloids.

The Deltoid

A hypocycloid is a plane curve which is the locus of a fixed point P on the perimeter of a small circle as it rolls without slipping within a larger circle as in figure 13.4. The ratio of the radius of the larger circle to the radius of the smaller circle determines the number of cusps of the curve and if this ratio is an integer (and therefore the ratio of the circumferences is an integer) as the small circle rotates to cover the circumference of the larger circle once, there will be a precise number of cusps generated. For example, this occurs in figure 13.5, where the hypocycloid has three cusps, with $a/b = 3$; this has been given the special

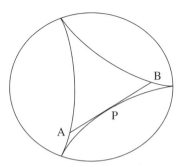

Figure 13.6.

name *deltoid*, and was first studied in 1745 by the incomparable Leonard Euler in connection with caustics.

The standard parametric form of the deltoid is

$$x = 2b \cos \theta + b \cos 2\theta,$$
$$y = 2b \sin \theta - b \sin 2\theta$$

for $0 \leqslant \theta \leqslant 2\pi$.

And from this all of its properties can be deduced, the two of which we need are

- the length of the tangent contained within the curve is constant and equal to $4b$;
- the area of the deltoid is $(a - b)(a - 2b)\pi$.

Figure 13.6 shows the tangent to the deltoid at a point P intersecting the curve at the two points A and B. If we force the tangent to be of length 1, it must be that $4b = 1$ and since $a/b = 3$ we have $a = \frac{3}{4}$ and $b = \frac{1}{4}$. Our needle is AB and we can reverse it within the deltoid by letting A trace the curve while keeping AB as a tangent to it; this guarantees that B is also on the curve, and as A moves from a cusp to the midpoint of the opposite side so AB inevitably reverses.

The area of the deltoid in which this reversal has been achieved is

$$\left(\frac{3}{4} - \frac{1}{4}\right)\left(\frac{3}{4} - 2 \times \frac{1}{4}\right)\pi = \frac{\pi}{8} = 0.3926 \cdots < \frac{1}{\sqrt{3}}$$

and, as Birkoff mentioned,

$$\frac{\pi}{8} = \frac{1}{2} \times \frac{\pi}{4}.$$

A Russian Solution from England

It was another émigré, this time from Russia, who continues the story. Abram Besicovitch had endured the devastating repercussions of the civil war that had erupted in Russia in 1917 as the Tzar's White Army and the Bolshevik Red Army repeatedly swapped control of the city of Perm, whose university boasted Vinogradov, Friedman and himself as professors. In 1920 a return to the devastated St Petersburg (now renamed Petrograd and where he had studied as a young man under Markov) with its vastly diminished university preceded his illicit escape in 1924— again to Copenhagen and again with the considerable and influential help of Bohr. From Copenhagen he went to Liverpool for a year (this time with the help of the great Hardy) and then to Cambridge, where he remained for the rest of his life. It is an eloquent measure of his quality that in 1950 he succeeded Littlewood to the prestigious Rouse Ball chair, which he occupied until his retirement in 1958. But we must return to war-torn Perm in 1917, when Besicovitch was working on a problem in Riemann integration (reproduced in the box below), which he reduced to the existence of planar sets of measure 0 which contain a unit line segment in each direction.

Given a Riemann-integrable function f on $\mathbb{R}^2$, must there exist a rectangular coordinate system (x, y) such that $f(x, y)$ is Riemann integrable as a function of x for each y, and that the two-dimensional integral of f is equal to the iterated integral $\iint f(x, y) \, dx \, dy$?

The intricacies of the problem's statement and the reasoning behind his reduction of it is of no concern to us here, it is only important to our story that he constructed such a set, the detail of which he published in 1919 in a Russian journal (Sur deux questions d'intégrabilité des fonctions, 1919, *Journal of the Society of Physics and Mathematics* 2:105-23). Unfortunately, the civil war and the ensuing international blockade ensured there was hardly any communication between Russia and the rest of the world at the time: Besicovitch had not heard of Kayeka and assuredly not of his problem. It was only after his escape to the West that he became aware of the problem, and possibly from Birkoff's 1925 book, mentioned earlier. An adaptation of his 1919 proof provided the astonishing answer to the Kakeya problem, which we restate below:

> In the class of all figures in which a segment of length 1 unit can be turned through 180°, remaining always within the figure, which one has the smallest area?

Besicovitch's answer is: *there isn't one.*

Kayeka would probably not have been too devastated had someone discovered another figure, perhaps esoteric, of area less than $\frac{1}{8}\pi$ in which the rotation could be accomplished but the result in Besicovitch's 1928 paper (A. S. Besicovitch, On Kakeya's problem and a similar one, 1928, *Mathematische Zeitschrift* 27:312-20) was altogether different. In short, he showed that there is no figure of smallest area; the task can be accomplished within any area we care to prescribe, no matter how small.

It has been said of Kayeka that he would jokingly think of the needle as a lance wielded by the ancient Japanese knights, the Samurai, commenting:

> The Samurai had a lance for protection, which he needed to be able to use freely in any room no matter what the size— even if the size of a lavatory.

Besicovitch's result had made the smallest room in the house a whole lot smaller. We will consider a version of his reasoning.

The Besicovitch Set

First, a Besicovitch set is a subset of the plane which contains a needle in every direction. (In fact, the idea easily extends to higher dimensions, a generalization which attracts extremely

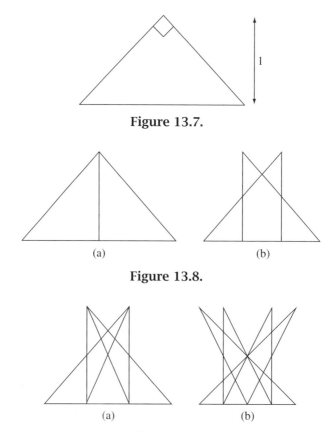

Figure 13.7.

(a)   (b)

Figure 13.8.

(a)   (b)

Figure 13.9.

difficult questions: for example, with a careful definition of the word, what is the minimum dimension of such a set?) Our interest is with such a planar set which has an arbitrarily small area. Besicovitch's original construction has undergone numerous improvements by him and others; we will consider one of the more modern approaches and start with a right-angled isosceles triangle of height 1 unit, as shown in figure 13.7.

We imagine the needle hinged at the right angle, sweeping out 90° as it swings anticlockwise from the left side of the triangle to the right: that the height of the triangle is 1 unit ensures that the needle when vertical is fully contained within the triangle. The figure contains all directions throughout the 90°.

Now we begin a sequence of constructions, the first of which is to cut the triangle along its median, as shown in figure 13.8(a), to make two smaller triangles and then slide the right of the two a distance to the left, as shown in figure 13.8(b). This new figure has the following properties:

- We can fancifully (and usefully) think of it as a 'batman cloak'.
- It can be thought of as a single triangle similar to the original (the body and triangular 'wings'), together with two 'ears'. (The similarity is caused by the two base angles of the triangles being correspondingly equal and so the top angles must be the same.)
- The area of the figure is strictly less than that of the original.
- When thought of as two overlapping triangles, one side of one triangle is parallel to one side of the other triangle. (At this stage these are the two vertical sides but, more generally, they will be the 'cut line(s)' along the median.)

The needle hinged at the top vertex of the left-hand triangle, lying along its left side, can be made to sweep out anticlockwise the first $45°$ of the $90°$ until it is vertical and lies along the right side of the triangle. Then it can be translated to the left to the top of the right-hand triangle and anticlockwise rotated by the second $45°$: this means that the figure also contains every direction over the $90°$ interval—and has a smaller area than the original figure, although we need extra area for that translation.

The process continues as follows.

Take each overlapping triangle and divide each in two by drawing the medians, as shown in figure 13.9(a). Translate the right part of the right triangle to the left and the left part of the left triangle to the right, as shown in figure 13.9(b). Now there are four triangles with which to repeat the procedure, always translating the right parts of the right triangles to the left and the left parts of the left triangles to the right. Figure 13.10 shows the next two stages of the procedure.

After each stage the figure is becoming ever more complicated but, because of the overlap, ever smaller in area, and it continues to contain all directions in the $90°$. To be clear about this, the left side of the original triangle remains in the figure and if

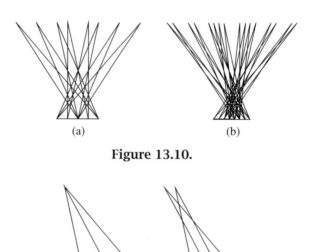

Figure 13.10.

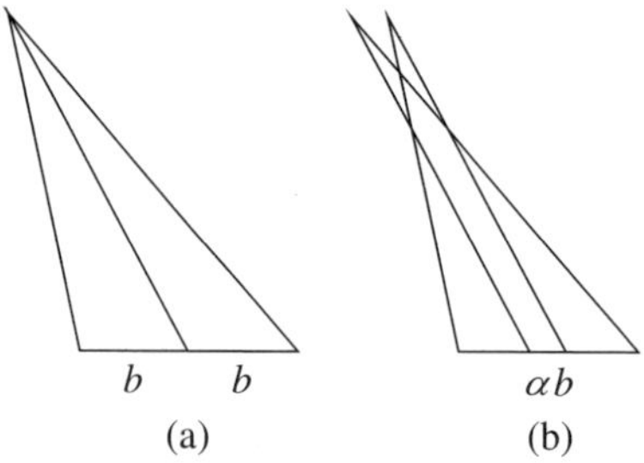

Figure 13.11.

the needle starts there we can rotate it anticlockwise by a small
number of degrees until it reaches the right side of that first
triangle, which has a matching parallel side in another triangle.
Translate the needle to that triangle and continue the rotation:
repeat the process for each small triangle and we will eventually
reach the last one, the right side of which is the right side of the
original triangle: having done so, we will have rotated the needle through 90°, if somewhat exotically. The figure really does
contain all directions in that 90° interval and assuredly the area
of it is strictly decreasing and it is a reasonable assumption that
it is bounded below by zero. (Technically, the limiting figure is
compact and has Lebesgue measure zero.)

A Calculation

For the sake of mathematical comfort, we will take a few lines to
show that the area of such a figure can be made to decrease to
zero.

  If we take a typical triangular region of vertical height 1

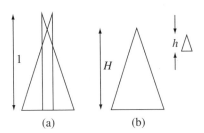

Figure 13.12.

and divide it by its median, we will have something like figure 13.11(a) and translating the right triangle to the left by a distance αb (where $0 < \alpha < 1$) brings about figure 13.11(b). We wish to calculate the area of the slanted 'batman cloak' consisting of the triangle (similar to the original) and the two 'ears'. Since shearing preserves area we can perform a horizontal shear with the base contained within the invariant line and which makes the two parallel sides vertical, as shown in figure 13.12(a).

If the triangle forming the cloak and the head has height H and that forming the head itself has height h, using similar triangles we have from figure 13.12 that

$$\frac{1}{2b} = \frac{H}{2b - \alpha b} = \frac{h}{\alpha b}$$

and this makes $H = \frac{1}{2}(2 - \alpha)$ and $h = \frac{1}{2}\alpha$ and so the area of the figure is

$$A = \frac{1}{2}(2b - \alpha b) \times H + 2 \times \frac{1}{2} \times 2h \times \frac{1}{2} \times \alpha b$$
$$= \frac{1}{4}(2 - \alpha)^2 b + \frac{1}{2}\alpha^2 b$$
$$= \frac{1}{4}(4 - 4\alpha + \alpha^2 + 2\alpha^2)b$$
$$= \frac{1}{4}(3\alpha^2 - 4\alpha + 4)b$$

and this, of course, is the area of the original slanted figure.

Since the area of the original triangle is $\frac{1}{2} \times 2b \times 1 = b$ we see that the ratio of the areas of the figure to the original triangle is $r = \frac{1}{4}(3\alpha^2 - 4\alpha + 4)$, which is reassuringly at most 1, as we can see from figure 13.13. Each repetition of the process doubles the number of triangles and creates a 'batman' figure, the

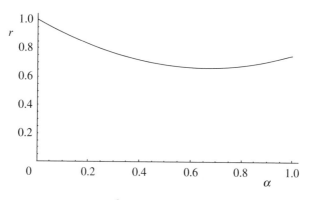

Figure 13.13.

area of which we have calculated. The area of the whole figure is strictly less than the sum of the areas of the 'batman' figures, since they are themselves made to overlap. We can, therefore, form an upper bound on the area of the figure after n repetitions by realizing that the base length satisfies

$$b_n = \alpha b_{n-1}, \quad n = 2, 3, \ldots,$$
$$b_1 = b$$

for each of the 2^{n-1} triangles and so the upper estimate of the area is

$$A_n = 2^{n-1} \times \tfrac{1}{4}(3\alpha^2 - 4\alpha + 4)b_n$$
$$= 2^{n-1} \times \tfrac{1}{4}(3\alpha^2 - 4\alpha + 4) \times (\alpha^{n-1}b)$$
$$= (2\alpha)^{n-1} \times \tfrac{1}{4}(3\alpha^2 - 4\alpha + 4)b.$$

Provided $2\alpha < 1$ as $n \to \infty$ it must be that $A_n \to 0$ and so the area of the figure must do so too.

Here we have the essence of Besicovitch's 1917 result, but to solve the Kayeka problem that needle needs to be rotated within the figure to achieve each of the directions and those translations from one parallel side to the next require us to move outside it and inevitably increase the area required to achieve our purpose. We will now consider the amount by which we need to increase the area.

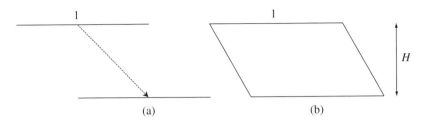

Figure 13.14.

Clever Translations

Figure 13.14(a) represents a translation of a unit needle by a vertical distance H. In undergoing the translation the needle will have traversed the area of the parallelogram shown in figure 13.14(b) and that area is $1 \times H = H$, which can be arbitrarily big.

Now let us consider an alternative approach to the translation. Suppose that we start with the needle and translate it in the direction of its length to the right by some distance R, rotate it anticlockwise about its centre by some angle (less than $\frac{1}{2}\pi$), translate the rotated needle downwards in the direction of its length and finally rotate the needle about its centre by the same angle clockwise. The effect of all of this is that a translation has been effected, as shown in a special case in figure 13.15.

We only need the two angles of rotation to be the same for the needle to be translated as desired, but let us suppose that the two translations described above are themselves the same distance R and that the angle of rotation is $1/R$ radians. Since we need $1/R < \frac{1}{2}\pi$, it must be that

$$R > \frac{2}{\pi} = 0.6366\ldots.$$

Using the formula for the area of a sector and this method, the total area traversed in achieving the translation is

$$2 \times \left(2 \times \frac{1}{2} \times \left(\frac{1}{2}\right)^2 \times \frac{1}{R}\right) = \frac{1}{2R}.$$

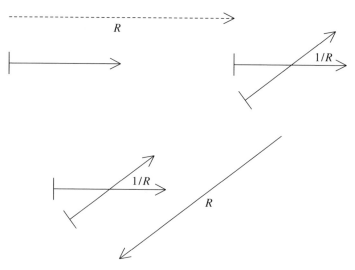

Figure 13.15.

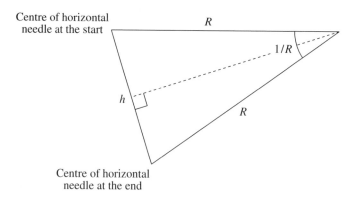

Figure 13.16.

Again, translating in the direction of the needle takes up no area. Now let us compare this expression with the area of the parallelogram which would have to be traversed to achieve the same translation.

If we form an isosceles triangle by joining the midpoints of the horizontal needle positions, we arrive at figure 13.16, where h is the distance the needle moves in the direction of motion. Elementary trigonometry shows that $h = 2 \times R \times \sin(1/2R)$.

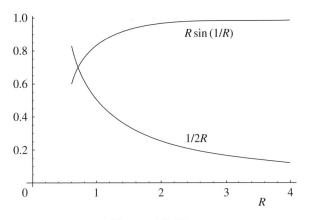

Figure 13.17.

The horizontal shift of the needle is then

$$h \times \sin\left(\frac{1}{2R}\right) = 2R \sin^2\left(\frac{1}{2R}\right)$$

and the vertical shift is

$$h \times \cos\left(\frac{1}{2R}\right) = 2 \times R \times \sin\left(\frac{1}{2R}\right) \times \cos\left(\frac{1}{2R}\right) = R\sin\left(\frac{1}{R}\right),$$

which makes the area of the parallelogram traversed by the needle

$$1 \times R\sin\left(\frac{1}{R}\right) = R\sin\left(\frac{1}{R}\right).$$

In summary, by the rotation method, we achieve the same trans-lation as a shift to the right by $2R\sin^2(1/2R)$ and down by $R\sin(1/R)$ but traverse an area of $1/2R$ rather than $R\sin(1/R)$.

Figure 13.17 shows a combined plot of $R\sin(1/R)$ and $1/2R$, which intersect at $R = 0.712\,04\ldots$, and makes clear that, for $R > 0.712\,04\ldots$,

$$\frac{1}{2R} < R\sin\left(\frac{1}{R}\right).$$

We can achieve a translation with an increasingly significant sav-ing of area as R increases if we adopt this more complicated

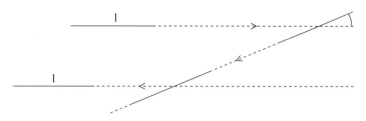

Figure 13.18.

technique. That said, the translation is special since

$$\lim_{R \to \infty} \left\{ 2R \sin^2 \left(\frac{1}{2R} \right) \right\} = 0 \quad \text{and} \quad \lim_{R \to \infty} \left\{ R \sin \left(\frac{1}{R} \right) \right\} = 1$$

means that the process progressively tends to translate the needle vertically down by 1 unit.

This special case of achieving a translation using an arbitrarily small area points the way to the final step needed to solve the Kayeka problem. The Besicovitch set comprises many triangles, one of which (T_L) originated from the left side of the original triangle and another (T_R) from the right side. To alter the direction of the needle by $90°$ we need to move it from the left side of (T_L) to the right side of (T_R) using the rest of the triangles as stepping stones. Movement within any triangle is achieved by anticlockwise rotation about its top vertex and movement between triangles by translation from the right side of one of them to the parallel left side of another: these translations can be made within an arbitrarily small area by adapting the above procedure, choosing two points A and B on the extended lines of the needle and its target side sufficiently far apart so that the rotation of the needle can be as small as required. Figure 13.18. represents this procedure.

The Kakeya Problem Solved

To solve the Kakeya problem we need to adopt the following strategy:

(i) Let the needle lie on the left side of T_L with one end at the upper vertex.

(ii) Rotate the needle anticlockwise about the upper vertex until it lies along the right side of T_L.

(iii) Locate the triangle with a side parallel to this side.

(iv) Perform the translation procedure to bring the needle to the left side of that triangle.

(v) Continue until the needle lies on the right side of T_R.

Adding the arbitrarily small area of the Besicovitch set to the sum of the arbitrarily small areas needed to perform the rotations succeeds in rotating the needle through 90° in an arbitrarily small area; attach another right-angled triangle and the needle can be rotated through 180° and translated along its own line back to its original position, in reverse order. Problem solved!

Littlewood, in his famous *Miscellany*, comments of his successor that in establishing Kakeya's result Besicovitch established the first of two counterintuitive phenomena. The second was in 1947 when he rediscovered a proof of *Crum's problem*, which asks for the maximum number of convex polyhedra for which every pair has at least part of their faces in common: the answer in two dimensions is 4 and it was thought that for three dimensions it would be 10 or 12. In fact for three dimensions Besicovitch (and Tietze before him) proved that the answer is that there is no limit to the number of them. To one reasonable question Besicovitch found the unreasonable answer to be zero and to another the equally unreasonable answer infinity.

So, is the problem completely solved? The answer is *no*, even for two dimensions. Moving from the equilateral triangle to the deltoid sacrifices convexity and moving from the deltoid to Besicovitch sets sacrifices another important topological property: *simple connectedness*. (This important term distinguishes between areas in which simple closed curves can be shrunk to a point and those for which this is not always possible; for example, a standard jam doughnut has no hole and so is simply connected, whereas the ring doughnut with the hole is not.) The deltoid is obviously simply connected and it is hardly surprising that Besicovitch sets are not. Is there a simply connected region with an area less than $\frac{1}{8}\pi$? No one knows. A move to higher dimensions creates enormous difficulties which, at the time of writing, remain unresolved. Over to the reader!

And is it more than a puzzle with a clever and surprising solution? We will leave it to a recent Field's Medallist to give his view:

> At first glance, Kakeya's problem and Besicovitch's resolution appear to be little more than mathematical curiosities. However, in the last three decades it has gradually been realized that this type of problem is connected to many other, seemingly unrelated, problems in number theory, geometric combinatorics, arithmetic combinatorics, oscillatory integrals, and even the analysis of dispersive and wave equations.

The words are those of Terence Tao, who is considered a world authority on the subject.

With these words in mind we will leave Kakeya's problem with a final extract from Birkoff's book:

> Since all physical science depends upon the foundation of mathematical truth, of which the discovery has been hastened by intellectual curiosity, a high place can be granted such curiosity for its proved value to mankind.

Chapter 14

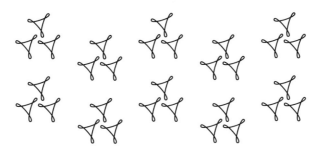

THE BEST CHOICE

My advice to you is to get married. If you find a good wife, you'll be happy; if not, you'll become a philosopher.

Socrates

Natural Selection

The German mathematician and astronomer Johannes Kepler was married twice, once in 1597 to Barbara Muhleck and, after her death from cholera, to Susanna Reuttinger in 1613: the first marriage was arranged by friends and matchmakers, the second by Kepler himself, who had assessed eleven possible candidates. Since he pondered the relative merits of the individuals for nearly two years, weighing parental standing, dowry size and conflicting advice each against the other he must be judged as a careful suitor. Eventually his decision was made and he explained his ultimate choice to one Baron Strahlendorf in a letter of 23 October 1613 of some dozen pages: in it he wrote that the lady had 'won me over with love, humble loyalty, economy of household, diligence, and the love she gave the stepchildren' (of which there were three, not counting a child of his late wife from a previous marriage) and that 'God had led him to choose the fifth'.

Here we are concerned with a very particular selection process, its rather surprising resolution and a currently inexplicable connection with the convergents of the partial fraction form of $1/e$.

The problem is most commonly expressed in terms of interviewing a list of individuals in order to select the best of them and has been given names such as the *secretary problem*, the *sultan's dowry problem* and, most appropriately for Kepler, the *fussy suitor problem*. In its generic form there is a collection of size n and it is desired to choose the best individual from it. In a random order, each individual is assessed and if rejected plays no further part; when one individual is selected the process ends with that selection; the hoped-for best from the list. If all are rejected up to the last, that last individual must be chosen.

The question is: what strategy could be adopted to maximize the chances of picking the best individual?

For example, if there is to be a 50% chance of choosing the best individual, it is intuitively reasonable that half of them should be assessed and the best one of these chosen; if the best candidate is in that half, that candidate will be chosen, and the probability of that happening is $\frac{1}{2}$. So for a probability of $\frac{1}{2}$ of choosing the best candidate there seems to have to be $\frac{1}{2}n$ interviews. We will look more deeply.

The origins of the problem are somewhat obscure but it may be reasonably argued that the distinguished and most prolific English mathematician, Arthur Cayley, may have been the first to articulate a version of it. Of the 966 papers attributed to him, paper 705 contains 50 pages of problems and solutions that he had submitted to the *Educational Times* between 1871 and 1894. One of the 1875 submissions was the following:

> 4528. (Proposed by Professor Cayley) A lottery is arranged as follows: There are n tickets representing $a, b, c, \ldots$ pounds respectively. A person draws once; looks at his ticket; and if he pleases, draws again (out of the remaining $n - 1$ tickets);

and so on, drawing in all not more than k times; and he receives the value of the last ticket drawn. Supposing that he regulates his drawings in the manner most advantageous to him according to the theory of probabilities, what is the value of his expectation?

The similarity to the generic problem is clear enough, but it is not precisely the same as it; here the *payoff* is whatever number of pounds is associated with the selected ticket yet with the generic example it is 1 or 0, depending on whether the best candidate is picked or not. As the years have passed the variants have grown in number and complexity until we have reached the stage at which the literature about it and its relatives is quite simply vast. This is a view borne out in Thomas S. Ferguson's comprehensive article 'Who solved the secretary problem?' (1989, *Statistical Science* 4(3):282–89), where he wrote:

> Since that time the problem has been extended and generalized in many different directions, so that now one can say that it constitutes a 'field' of study within mathematics-probability-optimization. One can see from the review paper by Freeman (1983) how extensive and vast the field has become; moreover, the field has continued its exponential growth in the years since that paper appeared.

The reference is to the article by P. R. Freeman 'The secretary problem and its extensions: a review' (1983, *International Statistical Review* 51(2):189–206) in which 'all published work to date on the problem and its extensions is reviewed'.

Strategy

The strategy we will consider for the interviewer is the following. *Interview and discard the first r candidates on the list and then choose the first candidate better than the best reject.*

The big question is, given that this is a sensible strategy at all, *what is this value of r?*

We can analyse the strategy in terms of the parameter r in the following way.

Suppose that the best individual is B. If this best individual happens to be in the $(r+1)$th position, we will choose B for certain; this happens with probability $1/n$. Now suppose that B is in the $(r+2)$th position, then, if the occupant of the $(r+1)$th position is the best yet, we will fail in our goal of choosing B, otherwise we will still choose B; the alternative is that the occupant is not the best yet, which means that the best yet among the first $r+1$ choices lies among the first r of them; this occurs with probability $r/(r+1)$. We do need B to be in the $(r+2)$th position, and this happens with that same probability of $1/n$, so the total probability of success in this case is

$$\frac{1}{n} \times \frac{r}{r+1}.$$

Now the process continues, supposing that B is in the $(r+3)$, $(r+4), \ldots, n$th position, giving the probabilities of success as

$$\frac{1}{n} \times \frac{r}{r+2}, \ \frac{1}{n} \times \frac{r}{r+3}, \ \ldots, \ \frac{1}{n} \times \frac{r}{n-1}.$$

Adding these up then gives the probability of choosing the best candidate as

$$P(n,r) = \frac{1}{n}\left(1 + \frac{r}{r+1} + \frac{r}{r+2} + \frac{r}{r+3} + \cdots + \frac{r}{n-1}\right).$$

Table 14.1 shows the values of $P(n,r)$ for $n = 1,2,3,\ldots,11$ and reveals the fact that, had Kepler used this technique (which undoubtedly he did not), he should have chosen the fourth suitor, where $P(11,4) = 0.398\,413$, rather than the fifth, where $P(11,5) = 0.384\,38$. Actually, history reveals that his friends intervened here, as Susanna Reuttinger was an orphan who lacked dowry and social standing, and they persuaded him to offer his hand to none other than candidate number 5: this he did—and was rejected. So, it was to be number 4, but all was well as they enjoyed a happy and fruitful time together, parenting six more children.

Computing the optimal r for a given n is so far a simple but unsatisfactory numerical exercise and we can make pleasing headway by once again using the harmonic series, which we

Table 14.1.

n	r										
	1	2	3	4	5	6	7	8	9	10	11
1	1										
2	0.5	0.5									
3	0.5	0.3333	0.3333								
4	0.458333	0.416667	0.25	0.25							
5	0.416667	0.433333	0.35	0.2	0.2						
6	0.380556	0.427778	0.391667	0.3	0.166667	0.166667					
7	0.35	0.414286	0.407143	0.352381	0.261905	0.142857	0.142857				
8	0.324107	0.398214	0.409821	0.379762	0.318452	0.232143	0.125	0.125			
9	0.301984	0.381746	0.405952	0.393122	0.352513	0.289683	0.208333	0.111111	0.111111		
10	0.282897	0.365794	0.39869	0.398254	0.372817	0.327381	0.265278	0.188889	0.1	0.1	
11	0.26627	0.350722	0.389719	0.398413	0.38438	0.352165	0.304798	0.244444	0.172727	0.0909091	0.0909091

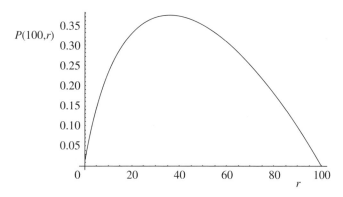

Figure 14.1.

mentioned in chapter 11. Recall that it is defined by

$$H_n = 1 + \frac{1}{2} + \frac{1}{3} + \frac{1}{4} + \cdots + \frac{1}{n}$$

and that the infinite series diverges. In terms of it, the probability becomes

$$P(n,r) = \frac{1}{n}(1 + r(H_{n-1} - H_r)).$$

Now we can replace the harmonic series itself by its logarithmic approximation $H_n \approx \ln n + y$, where $y = 0.577\,216\ldots$ is Euler's constant, which we have already seen on page 101.

The expression then becomes

$$P(n,r) \approx \frac{1}{n}\{1 + r([\ln(n-1) + y] - [\ln r + y])\}$$

$$= \frac{1}{n}\left(1 + r\ln\frac{(n-1)}{r}\right).$$

Figure 14.1 shows a plot of $P(100,r)$ against r, which typifies the general behaviour of the function.

We seek the value of r which maximizes the function for a given value of n and, if we treat r as a continuous variable, we can use calculus to look at the growth of the function and arrive at the expression

$$\frac{\mathrm{d}P(n,r)}{\mathrm{d}r} = \frac{1}{n}\left(\ln\frac{n-1}{r} - r \times \frac{1}{r}\right) = \frac{1}{n}\left(\ln\frac{n-1}{r} - 1\right).$$

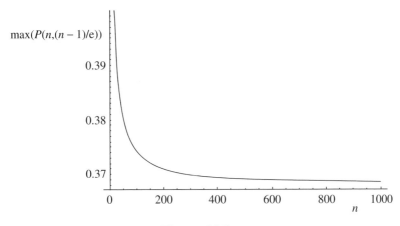

Figure 14.2.

For a maximum we require

$$\frac{\mathrm{d}P(n,r)}{\mathrm{d}r} = 0,$$

which means that

$$\ln \frac{n-1}{r} = 1 \quad \text{and} \quad \frac{n-1}{r} = \mathrm{e}.$$

So the maximum is achieved at $r = (n-1)/\mathrm{e}$, or, to be precise, at an integer either side of this value. The maximum value of the probability is then

$$P\left(n, \frac{n-1}{\mathrm{e}}\right) = \frac{1}{n}\left(1 + \frac{n-1}{\mathrm{e}}\ln\mathrm{e}\right)$$

$$= \frac{1}{n}\left(1 + \frac{n-1}{\mathrm{e}}\right) \xrightarrow[n\text{ large}]{} \frac{1}{n} \times \frac{n}{\mathrm{e}} = \frac{1}{\mathrm{e}} = 0.3678\ldots.$$

Figure 14.2 shows this asymptotic behaviour of the maximum value of the function

$$P\left(n, \frac{n-1}{\mathrm{e}}\right)$$

as a function of n and makes clear that it settles to the constant value of $1/\mathrm{e}$.

In summary, if about the first 37% of the individuals are assessed and rejected, then there is about a 37% chance of choosing the best candidate.

It is that number $1/e$ that we are interested in; to be accurate we are interested in particularly important fractional approximations to it.

Continued Fractions

A continued fraction is an expression of the form

$$a_0 + \cfrac{1}{a_1 + \cfrac{1}{a_2 + \cfrac{1}{a_3 + \cfrac{1}{a_4 + \cdots}}}},$$

where a_0 is an integer (possibly negative or zero) and $a_1, a_2, \ldots$ are nonzero positive integers; the expression could be finite or it could go on forever. Standard fractional notation is cumbersome and has given way to the alternative $[a_0; a_1, a_2, \ldots]$, in which the semicolon separates the number's integer from its fractional part and the commas separate what are known as its 'partial quotients'.

For example,

$$3 + \cfrac{1}{2 + \cfrac{1}{5 + \frac{1}{4}}} = 3 + \cfrac{1}{2 + \cfrac{1}{\left(\frac{21}{4}\right)}} = 3 + \cfrac{1}{2 + \frac{4}{21}}$$

$$= 3 + \cfrac{1}{\left(\frac{46}{21}\right)} = 3 + \frac{21}{46} = \frac{159}{46}$$

or, in a more compact notation, $[3; 2, 5, 4] = \frac{159}{46}$.

If we build up the expression one term at a time, we get

$$3 + \tfrac{1}{2} = \tfrac{7}{2} \quad \text{and} \quad 3 + \cfrac{1}{2 + \frac{1}{5}} = \frac{38}{11},$$

thereby generating the 'convergents' of the partial fraction. Put another way, $\frac{159}{46}$ is approximately $\frac{7}{2}$ and also $\frac{38}{11}$, with the latter the better approximation. Clearly, any finite continued fraction can be telescoped into an ordinary fraction in this way, with each of the convergents successively better approximants to that fraction. Converting an ordinary fraction to its continued form simply requires us to strip off the integer part, invert and repeat the process; for example,

$$\frac{18}{13} = 1 + \frac{5}{13} = 1 + \frac{1}{(\frac{13}{5})} = 1 + \frac{1}{2 + \frac{3}{5}} = 1 + \cfrac{1}{2 + \cfrac{1}{(\frac{5}{3})}}$$

$$= 1 + \cfrac{1}{2 + \cfrac{1}{1 + \frac{2}{3}}} = 1 + \cfrac{1}{2 + \cfrac{1}{1 + \cfrac{1}{(\frac{3}{2})}}} = 1 + \cfrac{1}{2 + \cfrac{1}{1 + \cfrac{1}{(1 + \frac{1}{2})}}}$$

or $[1; 2, 1, 1, 2]$, and, in the same way, $\frac{18}{13}$ is successively (and more accurately) approximated by $\frac{3}{2}$, $\frac{5}{3}$ and $\frac{7}{5}$.

The process of converting an irrational number to a continued fraction simply requires the decimal part to be dealt with in much the same way as a rational number. For example,

$$\pi = 3 + 0.141\,59\cdots = 3 + \frac{1}{7.062\,513\ldots}$$

$$= 3 + \cfrac{1}{7 + \cfrac{1}{15.996\,594\ldots}}$$

$$= 3 + \cfrac{1}{7 + \cfrac{1}{15 + \cfrac{1}{1.003\,417\ldots}}}$$

$$= 3 + \cfrac{1}{7 + \cfrac{1}{15 + \cfrac{1}{1 + \cfrac{1}{292 + 0.654\ldots}}}},$$

which continues as $\pi = [3; 7, 15, 1, 292, 1, 1, 1, 2, 1, 3, 1, 14, 2, 1, 1, 2, 2, 2, 2, 1, 84, \ldots]$, with initial convergents $\frac{22}{7}$, $\frac{333}{106}$, $\frac{355}{113}$ and

$\frac{103\,993}{33\,102}$: the first, of course, is the entirely familiar rational approximation to π.

It is also true that $\pi^4 = [97; 2, 2, 2, 2, 16539, 1, \dots]$, which makes the fifth convergent $\frac{3\,544\,4733}{363\,875}$ a particularly accurate rational approximation to π^4 (and therefore its fourth root a particularly accurate decimal approximation to π—differing in the thirteenth decimal place).

The continued fraction form for other numbers can be found in the same way and can reveal an otherwise hidden pattern. For example,

$$\sqrt{2} = [1; 2, 2, 2, 2, \dots] \quad \text{with convergents } \tfrac{3}{2}, \tfrac{7}{5}, \tfrac{10}{7}, \dots,$$
$$e = [2; 1, 2, 1, 1, 4, 1, 1, 6, 1, 1, 8, 1, 1, 10, 1, 1, 12, \dots]$$
$$\text{with convergents } \tfrac{5}{2}, \tfrac{8}{3}, \tfrac{11}{4}, \tfrac{19}{7}, \tfrac{73}{32}, \dots,$$

the 'Golden Ratio'

$$\varphi = \frac{1 + \sqrt{5}}{2}$$
$$= [1; 1, 1, 1, 1, \dots] \quad \text{with convergents } \tfrac{2}{1}, \tfrac{3}{2}, \tfrac{5}{3}, \tfrac{8}{5}, \dots.$$

This last example points us in the direction in which we wish to head. If we consider the numerator and denominator of the convergents of φ separately, they are consecutive terms in the Fibonacci sequence $1, 1, 2, 3, 5, 8, 13, \dots$. We will look at the consequences of separating the convergents of our key number $1/e$ in the same way.

The continued fraction form of $1/e$ is

$$\{0; 2, 1, 2, 1, 1, 4, 1, 1, 6, 1, 1, 8, 1, 1, 10, 1, 1, 12, 1, \dots\}$$

with convergents

$$\tfrac{1}{2}, \tfrac{1}{3}, \tfrac{3}{8}, \tfrac{4}{11}, \tfrac{7}{19}, \tfrac{32}{87}, \tfrac{39}{106}, \tfrac{71}{193}, \tfrac{465}{1264}, \tfrac{536}{1457}, \dots$$

The reader will discern a pattern beginning to emerge in the continued fraction form of the number, but it is with the successive fractional approximations that we are concerned.

Returning to our interview list and to table 14.1, we recall that, for a list of length 2, the optimal number to interview is 1, for

Table 14.2.

n	Optimal r	$\max P(n,r)$
2	1	0.500
3	1	0.500
8	3	0.409 8
11	4	0.398 4
19	7	0.385 0
87	32	0.371 5
106	39	0.370 9
193	71	0.369 5
1264	465	0.368 13
1457	536	0.368 10

a list of length 3 the optimal number to interview is 1, for a list of length 8 the optimal number to interview is 3, for the Keplerian list of length 11 the optimal number to interview is 4, etc. These pairs of numbers are the denominators and numerators respectively of the convergents of $1/e$.

Table 14.2 shows some more examples of the denominators (n) and numerators (r) of the convergents of $1/e$ providing the optimal combination. A sterner test is $n = 14\,665\,106$, which is the denominator of the twentieth convergent of $1/e$; its numerator is $5\,394\,991$, and that is precisely the optimal value of r. Put another way, we know that the optimal r for a given n satisfies the approximation $r \approx (1/e) \times n$; if n happens to be a denominator of a continued fraction approximant of $1/e$, then r is precisely the numerator of that fraction. Does this continue forever? We have no idea and, as far as we can tell, neither has anybody else. It is most peculiar.

Chapter 15

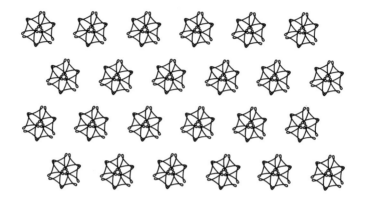

THE POWER OF POWERS

Sometimes it is useful to know how large your zero is.

Anon

Given its importance to the chapter, 'Some consequences of the irrationality of $\log_{10} 2$' could have been a reasonable alternative title. Yet the title stands, as the subject matter that follows concentrates on some rather surprising consequences relating to the decimal expansion of 2^n; we prove the elementary result that $\log_{10} 2$ is indeed irrational in the appendix (page 226), expending our efforts over the next few pages appealing to that result.

A Great Deal of Nothing

In a note to the journal *Mathematics of Computation* (E. and U. Karst, The first power of 2 with 8 consecutive zeroes, July 1964, 18(87):508) the authors provided what the note's title suggests:

Table 15.1.

Consecutive zeros	Power of 2
2	53
3	242
4	377
5	1 491
6	1 492
7	6 801
8	14 007

the first power of 2 which contains precisely eight consecutive zeros. They acknowledged that their work built on that of F. Gruenberger, who had computed the first power of 2 to contain 4, 5, 6 and 7 consecutive zeros, which, together with the Karsts' and the cases 2 and 3, are listed in table 15.1.

To be explicit, taking the first case,

$$2^{53} = 9\,007\,199\,254\,740\,992$$

is the first power of 2 to contain two consecutive zeros; the 4217 digits of $2^{14\,007}$ would take up too much space to write out, but the relevant part of the decimal expansion is

$$\cdots 6\,603\,000\,000\,003\,213\cdots.$$

The Karsts' IBM 1620 computer took 1 hour 18 minutes to find those eight consecutive zeros on 1 January 1964 and they reported that, on 1 May of the same year, they had tested up to the power 60 000 without finding the elusive nine-zero repetition; perhaps the interested reader might wish to use more modern technology to perform their own search. One thing is certain though: theoretically that search will not be in vain, since it is a fact that any number of repetitions is possible, 9, 900, 9000, or whatever number one wishes—although the numbers involved will assuredly be fantastically large.

To establish this peculiar fact we will look to the Elementary Problems and Solutions section of the *American Mathematical Monthly* (December 1963, 70(10):1101–2), wherein E. J. Burr (in particular) provided a succinct argument which we expand below.

First we need a result from the theory of rational approximation of irrational numbers, which is proved in the appendix (page 231):

> Given any irrational number λ and any positive integer k, there is a rational number m/n with $n \leqslant k$ such that $0 < \lambda - m/n < 1/nk$.

In particular, since $\log_{10} 2$ is irrational this result becomes

$$0 < \log_{10} 2 - \frac{m}{n} < \frac{1}{nk},$$
$$0 < n \log_{10} 2 - m < \frac{1}{k}$$

and, since 10^x is monotone increasing,

$$10^0 < 10^{n \log_{10} 2 - m} < 10^{1/k},$$
$$1 < 2^n \times 10^{-m} < 10^{1/k},$$
$$10^m < 2^n < 10^m \times 10^{1/k}.$$

Now take k large enough so that, for some chosen positive integer s, $10^{1/k} \leqslant 1 + 10^{-(s+1)}$, then $10^m < 2^n < 10^m \times (1 + 10^{-(s+1)})$, i.e. $10^m < 2^n < 10^m + 10^{m-s-1}$, which ensures that 2^n starts with a 1 followed by at least $m - (m - s - 1) - 1 = s$ consecutive zeros.

The Start of Something Big

The above proof locates the zeros after an initial digit 1 at the start of the decimal expansion yet we have seen evidence that the consecutive zeros can exist anywhere within the body of the number (and there are proofs for this too). This specialization suits our purpose though, since the main thrust of this chapter is that, no matter what sequence of whatever length of nonnegative integers we choose, there is at least one power of 2 which starts with that sequence.

If we start simply and require a power of 2 to start with the single digits $1, 2, 3, \ldots, 9$ in turn we arrive at table 15.2: notice

Table 15.2.

The first digit	The power of 2	2 to the power
1	4	16
2	1	2
3	5	32
4	2	4
5	9	512
6	6	64
7	46	$7.036 \cdots \times 10^{13}$
8	3	8
9	53	$9.0072 \cdots \times 10^{15}$

how hard we have to work for 7 and 9, the latter achieved by that same power to first contain two consecutive zeros.

If we become more ambitious and ask for the decimal expansion to start with the year of publication of this book, 2008, we require $2^{197} = 2008\ldots$. Higher ambition still yields the following sequence:

$$2^{47} = 14\ldots, \qquad 2^{243} = 141\ldots,$$
$$2^{6651} = 1414\ldots, \qquad 2^{35389} = 14142\ldots$$

and we are beginning to generate the most significant part of the decimal expansion of $\sqrt{2}$. If we tame the gargantuan number so generated by dividing it by the appropriate power of 10, we have

$$\frac{2^{35389}}{10^{\lceil 35389 \log 2 \rceil - 1}} = 1.41412 \cdots \sim \sqrt{2}.$$

If we wish to demand more and boast further decimal places we can have them—as many as we please, although the numbers involved will be vast. (Here, $\lceil x \rceil$ is the *ceiling* of x, discussed in the appendix (page 227).)

Moving to other attractive examples we also have

$$\frac{2^{51684}}{10^{\lceil 51684 \log 2 \rceil - 1}} = 2.7182 \cdots \sim e,$$
$$\frac{2^{55046}}{10^{\lceil 55046 \log 2 \rceil - 1}} = 3.1415 \cdots \sim \pi.$$

Of course these powers of 2 do make the numbers generated vastly big; to get some sort of idea *how* big, if we take a sheet of paper 0.1 mm thick and double it over 100 times(!) the thickness, 0.1×2^{100} mm, will be greater than the distance to the furthest galaxy.

Now we need to prove the fact that *any sequence of digits whatsoever can form the most significant digits of some power of 2* and to achieve this we will appeal to an argument provided by Yaglom and Yaglom in the first of their two books of mathematical problems (*Challenging Mathematical Problems with Elementary Solutions*, 1987, A. M. Yaglom and I. M. Yaglom, Dover), expanded by Ross Honsberger in his volume *Ingenuity in Mathematics* (*Mathematical Association of America*, 1970). We will expand further; the argument is indeed *elementary*, as it is *ingenious*—and one could also argue that it is extremely subtle and instructive. First, we will restate the proposition in an equivalent form.

The Restatement

If we consider the $\sqrt{2}$ example above, we can rephrase matters by asking the question: is there a positive integer n so that

$$14\,142 \cdots \leqslant 2^n < 14\,143\ldots?$$

If such an n exists, we will assuredly have 2^n beginning with the digits 14 142. Now replace the ellipses $\ldots$ by an appropriate power of 10 to get

$$14\,142 \times 10^k \leqslant 2^n < 14\,143 \times 10^k.$$

Generally, if we wish a power of 2 to start with the sequence M, we require positive integers k and n so that

$$M \times 10^k \leqslant 2^n < (M+1) \times 10^k$$

and applying the monotone increasing $\log_{10}$ function throughout gives

$$\log_{10}(M \times 10^k) \leqslant \log_{10} 2^n < \log_{10}((M+1) \times 10^k).$$

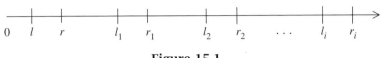

Figure 15.1.

Using the standard laws of logs the restatement is then the following.

For a given positive integer M we require positive integers k and n so that

$$k + \log_{10} M \leqslant n \log_{10} 2 < k + \log_{10}(M + 1).$$

The Proof

To show that such k and n exist we will need two results. An application of the pigeon-hole principle, discussed in the appendix (page 228) and a fact arising from the interrelationship between logs and the floor function (discussed in the appendix (page 230)):

$$\lfloor \log_{10}(M + 1) \rfloor = \lfloor \log_{10} M \rfloor: \quad M + 1 \text{ not a power of } 10,$$
$$\lfloor \log_{10}(M + 1) \rfloor = \lfloor \log_{10} M \rfloor + 1: \; M + 1 \text{ a power of } 10.$$

With these parts in place we can proceed to the very devious proof.

Write $l = \log_{10} M$ and $r = \log_{10}(M + 1)$ and so define the semi-open interval $[l, r)$ whose interval length is

$$r - l = \log_{10}(M + 1) - \log_{10} M = \log_{10}\left(\frac{M + 1}{M}\right)$$
$$= \log_{10}\left(1 + \frac{1}{M}\right) < \log_{10} 2 < 1$$

since $M \geqslant 1$.

Now translate copies of this interval (whose length is strictly less than 1) repeatedly to the right by 1 unit to get the infinite set of nonoverlapping, semi-open intervals $[l_i, r_i) = [i + l, i + r)$ for $i = 1, 2, 3, \ldots$, as shown in figure 15.1.

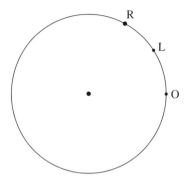

Figure 15.2.

Having done this, we will use the device of wrapping the whole of the infinite positive number line anticlockwise onto a circle of circumference 1 unit, which means that the line repeatedly and indefinitely overlaps. Numbers on the number line which differ by an integer will be mapped to the same point on the circle by this process and vice versa; in particular, all of the l_i will be mapped onto the same point L and all of the r_i onto the same point R. The situation is shown in figure 15.2, where the origin is at point O. It is these L and R that will provide the outside values of our double inequality.

Now we need to fit that multiple of $\log_{10} 2$ at the centre of the double inequality and to achieve that consider the infinite set of numbers on the number line

$$\log_{10} 2, \ 2\log_{10} 2, \ 3\log_{10} 2, \ \ldots, \ n\log_{10} 2, \ \ldots$$

and denote their images on the circle by $C_1, C_2, C_3, \ldots, C_n, \ldots$. These go round and round the circle in steps of arc length $\log_{10} 2$ and no two of them coincide, for if two did then

$$a\log_{10} 2 - b\log_{10} 2 = m$$

for some integer m, which would mean $\log_{10} 2 = m/(a-b)$ and would therefore be rational, and the irrationality of this number appears crucially for a second time.

We then have an infinite sequence of distinct points on the circle of (evidently) finite length and this means that there must

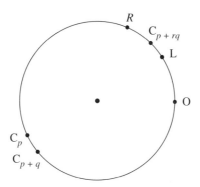

Figure 15.3.

exist a pair of them whose distance apart is smaller than any given number, which is simply an application of the pigeon-hole principle mentioned above. Let us then require two such to be closer together than the points L and R and label such a pair as C_pC_{p+q}, as shown in figure 15.3: if we define the function A to mean arc length, we have that $A(C_pC_{p+q}) < A(LR)$.

Now consider the infinite sequence of points C_p, C_{p+q}, C_{p+2q}, ..., C_{p+rq}, ... that correspond to the numbers $p \log_{10} 2$, $(p + q) \log_{10} 2$, $(p+2q) \log_{10} 2, \ldots, (p+rq) \log_{10} 2, \ldots$ on the number line; then

$$A(C_pC_{p+q}) = A(C_{p+q}C_{p+2q}) = A(C_{p+2q}C_{p+3q}) = \cdots$$
$$= A(C_{p+(r-1)q}C_{p+rq}) = \cdots = q \log_{10} 2 < A(LR)$$

since each adjacent pair is a distance $q \log_{10} 2$ apart on the circle.

Since this sequence of points is going around the circle in steps of length less than $A(LR)$ on each revolution, at least one of them must lie in the arc LR; let one such be C_{p+rq}, then

$$A(OL) \leqslant A(OC_{p+rq}) < A(OR).$$

Now we will convert this inequality of arc lengths to the corresponding inequality of the numbers on the number line, realizing that, if $x = OX$ on the number line, its image on the circle is the point X on it so that $A(OX) =$ the fractional part of $x = \{x\} = x - \lfloor x \rfloor$.

Apply this to the points $l = \mathrm{OL}$, $r = \mathrm{OR}$ and $(p + rq)\log_{10} 2 = \mathrm{OC}_{p+rq}$ on the number line to get

$$l - \lfloor l \rfloor \leqslant (p + rq)\log_{10} 2 - \lfloor (p + rq)\log_{10} 2 \rfloor < r - \lfloor r \rfloor,$$

$$\log_{10} M - \lfloor \log_{10} M \rfloor \leqslant (p + rq)\log_{10} 2 - \lfloor (p + rq)\log_{10} 2 \rfloor$$
$$< \log_{10}(M + 1) - \lfloor \log_{10}(M + 1) \rfloor,$$

$$\lfloor (p + rq)\log_{10} 2 \rfloor - \lfloor \log_{10} M \rfloor + \log_{10} M \leqslant (p + rq)\log_{10} 2$$
$$< \lfloor (p + rq)\log_{10} 2 \rfloor - \lfloor \log_{10}(M + 1) \rfloor + \log_{10}(M + 1).$$
$$(15.1)$$

Our k in the restatement of the result on page 181 is defined as

$$k = \lfloor (p + rq)\log_{10} 2 \rfloor - \lfloor \log_{10} M \rfloor$$

and to ensure that it is positive we choose an r so big that this is certain.

Finally, we need to deal with that term $\lfloor \log_{10}(M + 1) \rfloor$ on the right of this double inequality and do so by considering two cases.

$M + 1$ is not a power of 10.   Using the first result on page 181, $\lfloor \log_{10}(M + 1) \rfloor = \lfloor \log_{10} M \rfloor$ and the inequality can be written

$$\lfloor (p + rq)\log_{10} 2 \rfloor - \lfloor \log_{10} M \rfloor$$
$$+ \log_{10} M \leqslant (p + rq)\log_{10} 2$$
$$< \lfloor (p + rq)\log_{10} 2 \rfloor - \lfloor \log_{10} M \rfloor + \log_{10}(M + 1)$$

and so as

$$k + \log_{10} M \leqslant (p + rq)\log_{10} 2 < k + \log_{10}(M + 1),$$

which is exactly what we want.

$M + 1$ is a power of 10.   Since $\{x\} = x - \lfloor x \rfloor$ = fractional part of x, the middle part of equation (15.1) must be less than 1 and so the inequality can be rewritten

$$\log_{10} M - \lfloor \log_{10} M \rfloor \leqslant (p + rq)\log_{10} 2 - \lfloor (p + rq)\log_{10} 2 \rfloor < 1.$$

Now we use the second result on page 181 to get

$$\lfloor \log_{10}(M+1) \rfloor - \lfloor \log_{10} M \rfloor = 1$$

and so

$$\log_{10} M - \lfloor \log_{10} M \rfloor \leqslant (p+rq)\log_{10} 2 - \lfloor (p+rq)\log_{10} 2 \rfloor$$
$$< \lfloor \log_{10}(M+1) \rfloor - \lfloor \log_{10} M \rfloor.$$

But now $\log_{10}(M+1)$ is an integer and so

$$\log_{10}(M+1) = \lfloor \log_{10}(M+1) \rfloor,$$

which means that

$$\log_{10} M - \lfloor \log_{10} M \rfloor \leqslant (p+rq)\log_{10} 2 - \lfloor (p+rq)\log_{10} 2 \rfloor$$
$$< \log_{10}(M+1) - \lfloor \log_{10} M \rfloor$$

and

$$\lfloor (p+rq)\log_{10} 2 \rfloor - \lfloor \log_{10} M \rfloor + \log_{10} M \leqslant (p+rq)\log_{10} 2$$
$$< \lfloor (p+rq)\log_{10} 2 \rfloor - \lfloor \log_{10} M \rfloor + \log_{10}(M+1),$$

and again

$$k + \log_{10} M \leqslant (p+rq)\log_{10} 2 < k + \log_{10}(M+1).$$

Recall that the restatement of the result was that, for a given positive integer M, we require positive integers k and n so that

$$k + \log_{10} M \leqslant n\log_{10} 2 < k + \log_{10}(M+1)$$

is true: with $n = p + rq$ and $k = \lfloor (p+rq)\log_{10} 2 \rfloor - \lfloor \log_{10} M \rfloor$ we have precisely that!

Equal Distribution and Probabilities

If we look back to table 15.2, we can allow its scant information to guide us in addressing a reasonable question: *of all of the infinite possibilities, what proportion of the numbers 2^n start with each of the digits from 1 to 9?*

Perhaps it is natural to take one of two views:

- In the long run things will even out and so the proportion of first digits will be constant across the nine possibilities: the answer is $\frac{1}{9}$.
- I am not sure, but that behaviour of 7 and 9 is suspicious: possibly, for some reason, the other digits occur with equal likelihood and those two are special.

In fact, neither answer is correct and we will need that irrationality of $\log_{10} 2$ one last time to establish the truth of the matter— and also a result known as *Weyl's Equidistribution Theorem*, which is an important theorem in analytic number theory and was established by the eminent early twentieth-century mathematician Herman Weyl. In its original form it can be stated as follows:

> For any irrational number α, the sequence $\{\{n\alpha\} = n\alpha - \lfloor n\alpha \rfloor : n = 1, 2, 3, \ldots\}$ is equidistributed modulo 1.

A precise (and technical) definition is that the sequence $\{x_n : n = 1, 2, 3, \ldots\}$ is *equidistributed modulo 1* if for every interval $(a, b) \subset [0, 1]$

$$\lim_{n \to \infty} \frac{N[\{\{x_1\}, \{x_2\}, \{x_3\}, \ldots, \{x_n\}\} \cap (a, b)]}{n} = b - a.$$

In other words, in the long run the proportion of fractional parts of the x_n that fall in any subinterval is just the length of the subinterval. For example, since the interval $(0.6, 0.8)$ occupies 20% of the whole interval $(0, 1)$, the proportion of members of the sequence x_n whose fractional part falls between 0.6 and 0.8 will approach 0.2.

Note that the irrationality of α is crucial. Consider, for example, $\alpha = \frac{4}{9}$, then the sequence

$$\{\{n \times \tfrac{4}{9}\} : n = 1, 2, 3, \ldots\} = \{\tfrac{4}{9}, \tfrac{8}{9}, \tfrac{3}{9}, \tfrac{7}{9}, \tfrac{2}{9}, \tfrac{6}{9}, \tfrac{1}{9}, \tfrac{5}{9}, 0, \tfrac{4}{9}, \ldots\}$$

repeats, generating a set of nine fractions (including 0) appearing with equal likelihood, which evidently do not satisfy equal distribution. (In fact, it is easy to show that this finite repetition is necessary and sufficient for α to be rational.)

Now we will see how all this helps with the distribution of those first digits of 2^N.

We know that 2^N begins with a digit d if

$$d \times 10^n \leqslant 2^N < (d+1) \times 10^n.$$

From this we can find n in terms of N as follows:

$$d \leqslant \frac{2^N}{10^n} < (d+1).$$

Again, using the increasing monotonicity of $\log_{10}$,

$$\log_{10} d \leqslant \log_{10}\left(\frac{2^N}{10^n}\right) < \log_{10}(d+1),$$

$$0 \leqslant \log_{10} d \leqslant \log_{10}\left(\frac{2^N}{10^n}\right) < \log_{10}(d+1) \leqslant 1,$$

$$0 \leqslant \log_{10}\left(\frac{2^N}{10^n}\right) < 1,$$

$$\left\lfloor \log_{10}\left(\frac{2^N}{10^n}\right) \right\rfloor = 0,$$

$$\lfloor \log_{10} 2^N - n \rfloor = 0,$$

which means that the difference between the two is less than 1 and, since n is an integer, it must be that

$$n = \lfloor \log_{10} 2^N \rfloor.$$

Now we return to the original inequality to arrive at

$$\log_{10}(d \times 10^n) \leqslant \log_{10} 2^N < \log_{10}((d+1) \times 10^n),$$

$$\log_{10}(d \times 10^n) \leqslant \lfloor \log_{10} 2^N \rfloor + \{\log_{10} 2^N\}$$
$$< \log_{10}((d+1) \times 10^n),$$

$$\log_{10} d + n - \lfloor \log_{10} 2^N \rfloor \leqslant \{\log_{10} 2^N\}$$
$$< \log_{10}(d+1) + n - \lfloor \log_{10} 2^N \rfloor,$$

$$\log_{10} d \leqslant \{N \log_{10} 2\} < \log_{10}(d+1),$$

since $n = \lfloor \log_{10} 2^N \rfloor$.

Table 15.3.

d	Probability
1	0.301 03
2	0.176 09
3	0.124 93
4	0.096 91
5	0.079 18
6	0.066 94
7	0.057 99
8	0.051 15
9	0.045 75

Since by Weyl's Equidistribution Theorem $\{N \log_{10} 2\}$ is equidistributed modulo 1,

$$P[\log_{10} d \leqslant \{N \log_{10} 2\} < \log_{10}(d + 1)]$$
$$= \log_{10}(d + 1) - \log_{10} d = \log_{10}\left(\frac{d + 1}{d}\right) = \log_{10}\left(1 + \frac{1}{d}\right)$$

and so we have the remarkable fact that

$$P[\text{First digit of } 2^N = d] = \log_{10}\left(1 + \frac{1}{d}\right).$$

Table 15.3 shows these probabilities.

This done, the same argument applies for any set of digits M and so we have the general result that

$$P[2^N \text{ starts with the digit sequence } M] = \log_{10}\left(1 + \frac{1}{M}\right).$$

Reverting to our earlier examples:

$$P[2^N \text{ starts with 2008}] = \log_{10}\left(1 + \frac{1}{2008}\right)$$
$$= 0.000\,216\,2\ldots,$$
$$P[2^N \text{ starts with 14\,142}] = \log_{10}\left(1 + \frac{1}{14\,142}\right)$$
$$= 0.000\,030\,70\ldots,$$

$$P[2^N \text{ starts with } 27\,182] = \log_{10}\left(1 + \frac{1}{27\,182}\right)$$
$$= 0.000\,015\,976\ldots,$$

$$P[2^N \text{ starts with } 31\,415] = \log_{10}\left(1 + \frac{1}{31\,415}\right)$$
$$= 0.000\,013\,82\ldots.$$

Now look back over the arguments of this chapter and the reader will see that the number 2 is not intrinsic to the results; any number a would do provided that $\log_{10} a$ is irrational, which means any a that is not a rational power of 10. In particular, this means that this final section of the distribution of most significant digits is more general than at first it might seem—and far more general than one could ever imagine. And that fact takes us nicely to the next chapter.

Chapter 16

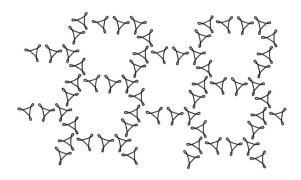

BENFORD'S LAW

Major paradoxes provide food for logical thought for
decades and sometimes centuries.

Nicholas Bourbaki

First Digits

At the end of the previous chapter, with the use of the potent
Weyl Equidistribution Theorem, we saw that the first digit of
powers of 2 are not distributed uniformly over $\{1, 2, 3, \ldots, 9\}$
but rather according to the law

$$P[\text{First digit of } 2^n = d] = \log_{10}\left(1 + \frac{1}{d}\right).$$

Moreover, we argued that the phenomenon exists equally with
2 replaced by any number that is not a rational power of 10.
Perhaps this behaviour is then a property of powers of integers,
but then consider the consumption (measured in kilowatt hours)
of the 1243 users of electricity of Honiara in the British Solomon

Table 16.1.

d	Logarithmic	Solomon
1	0.301 03	0.316
2	0.176 09	0.167
3	0.124 93	0.116
4	0.096 91	0.087
5	0.079 18	0.085
6	0.066 94	0.064
7	0.057 99	0.057
8	0.051 15	0.050
9	0.045 75	0.057

Islands in October 1969. Table 16.1 shows the proportion of use beginning with a first digit 1, etc., together with those numbers generated by the logarithmic formula.

Of course, the fit is not exact but it is markedly closer than the uniform $0.1111\ldots$: something seems to connect mathematical and electrical power. Something does: Benford's Law.

In 1881 the American mathematician and astronomer Simon Newcomb wrote to the *American Journal of Mathematics* (Note on the frequency of use of the different digits in natural numbers, 1881, 4(1):39-40) an article which began:

> That the ten digits do not occur with equal frequency must be evident to any one making much use of logarithmic tables, and noticing how much faster the first pages wear out than the last ones. The first significant figure is oftener 1 than any other digit, and the frequency diminishes up to 9.

This era, long before the invention of the electronic chip, depended on tables of logarithms for anything other than the simplest calculations; compiled into books these would have been a common sight in any mathematician's or scientist's place of work. Newcomb had noticed that the books of logarithms that he shared with other scientists showed greater signs of use at their beginning than they did at their end, but since logarithm tables were arranged in ascending numeric order, this suggested that more numbers with small rather than large first digits were being used for calculation. In the article he suggested an empirical law that the fraction of numbers that start with the digit d is not that intuitively reasonable $\frac{1}{9}$ but that remarkable $\log_{10}(1 + 1/d)$.

There was no rigorous justification provided and the idea languished in the mathematical shadows until 1938. It was then that Frank Benford, a physicist at G.E.C., published the paper 'The law of anomalous numbers' (*Proceedings of the American Philosophical Society*, 1938, 78:551–72). In it he had compiled a table of frequency of occurrence of first digits of 20 229 naturally occurring numbers, which is reproduced in table 16.2, and which he had extracted from a wide variety of sources ranging from articles in a selection of newspapers to the size of town populations; the logarithmic not the uniform law held the more convincingly. In particular the penultimate row, which averages the data, is a most excellent fit to the logarithmic model. The phenomenon has subsequently become known as *Benford's Law*.

Some Rationale

Two significant and reasonable observations have been made.

One, that if Benford's Law does hold, it must do so as an intrinsic property of the number systems we use. It must, for example, apply to the base 5 system of counting of the Arawaks of North America, the base 20 system of the Tamanas of the Orinoco and to the Babylonians with their base 60, as well as to the exotic Basque system, which uses base 10 up to 19, base 20 from 20 to 99 and then reverts to base 10. The law must surely be base independent.

The second is that changing the units of measurement must not change the frequency of first significant digits. Ralph A. Raimi, in his survey of progress on the matter (The first digit problem, *American Mathematical Monthly*, 1976, 83:521–38) wrote the following:

> Roger S. Pinkham attributing the basic idea to R. Hamming, put forward an invariance principle attached to another sort of probability model, sufficient to imply Benford's Law. If (say) a table of physical constants or of the surface areas

Table 16.2.

Title	First digit									Samples
	1	2	3	4	5	6	7	8	9	
Rivers, area	31.0	16.4	10.7	11.3	7.2	8.6	5.5	4.2	5.1	335
Population	33.9	20.4	14.2	8.1	7.2	6.2	4.1	3.7	2.2	3259
Physical constants	41.3	14.4	4.8	8.6	10.6	5.8	1.0	2.9	10.6	104
Numbers from newspaper articles	30.0	18.0	12.0	10.0	8.0	6.0	6.0	5.0	5.0	100
Specific heat	24.0	18.4	16.2	14.6	10.6	4.1	3.2	4.8	4.1	1389
Pressure	29.6	18.3	12.8	9.8	8.3	6.4	5.7	4.4	4.7	703
H.P. lost	30.0	18.4	11.9	10.8	8.1	7.0	5.1	5.1	3.6	690
Molecular weight	26.7	25.2	15.4	10.8	6.7	5.1	4.1	2.8	3.2	1800
Drainage	27.1	23.9	13.8	12.6	8.2	5.0	5.0	2.5	1.9	159
Atomic weight	47.2	18.7	5.5	4.4	6.6	4.4	3.3	4.4	5.5	91
$n^{-1}, n^{1/2}$	25.7	20.3	9.7	6.8	6.6	6.8	7.2	8.0	8.9	5000
Design	26.8	14.8	14.3	7.5	8.3	8.4	7.0	7.3	5.6	560
'Readers Digest' data	33.4	18.5	12.4	7.5	7.1	6.5	5.5	4.9	4.2	308
Cost data	32.4	18.8	10.1	10.1	9.8	5.5	4.7	5.5	3.1	741
X-ray volts	27.9	17.5	14.4	9.0	8.1	7.4	5.1	5.8	4.8	707
American league	32.7	17.6	12.6	9.8	7.4	6.4	4.9	5.6	3.0	1458
Blackbody	31.0	17.3	14.1	8.7	6.6	7.0	5.2	4.7	5.4	1165
Addresses	28.9	19.2	12.6	8.8	8.5	6.4	5.6	5.0	5.0	342
Mathematical constants	25.3	16.0	12.0	10.0	8.5	8.8	6.8	7.1	5.5	900
Death rate	27.0	18.6	15.7	9.4	6.7	6.5	7.2	4.8	4.1	418
Average	30.6	18.5	12.4	9.4	8.0	6.4	5.1	4.9	4.7	1011
Probable error (+ve/−ve)	0.8	0.4	0.4	0.3	0.2	0.2	0.2	0.2	0.3	

of a set of nations or lakes, is rewritten in another system of units of measurement, ergs for foot-pounds or acres for hectares, the result will be a rescaled table whose every entry is the same multiple of the corresponding entry in the original table. If the first digits of all the tables in the universe obey some fixed distribution law, Stigler's or Benford's or some other, that law must surely be independent of the system of units chosen, since God is not known to favour either the metric system or the English system. In other words, a universal first digit law, if it exists, must be scale-invariant.

Recalling our earlier example of electricity consumption, it should not matter whether kilowatt hours were used as a unit or any other alternative. Roger Pinkham, then a mathematician at Rutgers University in New Brunswick (New Jersey), had written a paper demonstrating that scale invariance implies Benford's Law (On the distribution of first significant digits, *Annals of Mathematical Statistics*, 1961, 32:1223–30).

In 1995, Theodore Hill of the Georgia Institute of Technology approached matters differently. He showed that, if distributions are selected at random and random samples are taken from each of these distributions, the significant-digit frequencies of the combined sample would converge to conform to Benford's Law, even though the individual distributions selected may not; a result consistent with that penultimate row of Benford's table (A statistical derivation of the significant-digit law, *Statistical Science*, 1995, 10:354–63). In a sense, Benford's Law is the distribution of distributions.

All of this said, many sets of numbers certainly do not obey Benford's Law: random numbers at one extreme and numbers that are governed by some other statistical distribution on the other, perhaps uniform or normal. It seems that, for data to conform to the law, they need just the right amount of structure.

An Argument

'Proving' Benford's Law is not like proving a standard mathematical theorem: even stating it properly is difficult, but we will

Table 16.3.

Interval	First significant digit after ×2
$[1, 1.5)$	2
$[1.5, 2)$	3
$[2, 2.5)$	4
$[2.5, 3)$	5
$[3, 3.5)$	6
$[3.5, 4)$	7
$[4, 4.5)$	8
$[4.5, 5)$	9
$[5, 10)$	1

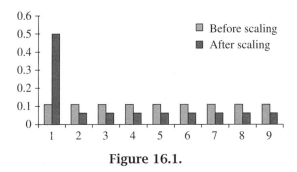

Figure 16.1.

approach it by following the scale-invariance property that it must, in all reason, adhere to.

A change of units is achieved by multiplying by some scaling number. Let us assume that the first significant digits of some measurable quantity are originally uniformly distributed and then let us suppose that we change the units by (say) multiplying everything by 2. If the first significant digit of the number in the original units is one of 5, 6, 7, 8, 9, the scaled number must begin with 1, otherwise, the behaviour is shown in table 16.3, which is displayed in the bar chart in figure 16.1. Even if the frequency of first significant digits was uniform before the scaling, it will not be afterwards and we are bound to conclude that equally likely digits are not scale invariant.

We can put forward a general argument using the standard theory of statistical distributions.

Recall that a continuous, nonnegative function $\varphi(x)$ is the probability density function of a continuous random variable X if $P(a \leqslant X \leqslant b) = \int_a^b \varphi(x)\,\mathrm{d}x$. Of course, we require that the total area under $\varphi(x)$ must be 1.

The cumulative density function $\Phi(x)$ is then defined by $\Phi(x) = P(X \leqslant x) = \int^x \varphi(t)\,\mathrm{d}t$ for an arbitrary lower limit, which means that

$$\frac{\mathrm{d}\Phi(x)}{\mathrm{d}x} = \varphi(x)$$

and

$$P(a \leqslant X \leqslant b) = \int_a^b \varphi(x)\,\mathrm{d}x = \Phi(b) - \Phi(a).$$

Now we will make precise the idea of a random variable being scale invariant and say that if it is so, the probabilities that it lies in some interval before and after scaling are the same. For our later convenience we will write the interval as $[\alpha, x]$ and the scale factor as $1/a$. Then scale invariance means

$$P(\alpha < X < x) = P\left(\alpha < \frac{1}{a}X < x\right) = P(a\alpha < X < ax).$$

This means that $\Phi(ax) - \Phi(a\alpha) = \Phi(x) - \Phi(\alpha)$ or $\Phi(ax) = \Phi(x) + K$ for all a.

So, assuming scale invariance, we have that $\Phi(ax) = \Phi(x) + K$ and differentiating both sides with respect to x gives $a\varphi(ax) = \varphi(x)$ and therefore $\varphi(ax) = (1/a)\varphi(x)$.

Now consider the random variable $Y = \log_b X$ with $\psi(y)$ and $\Psi(y)$ defined analogously. Then

$$\Psi(y) = P(Y \leqslant y) = P(\log_b X \leqslant y) = P(X \leqslant b^y)$$
$$= \Phi(b^y) = \Phi(x).$$

This means that

$$\psi(y) = \frac{\mathrm{d}}{\mathrm{d}y}\Psi(y) = \frac{\mathrm{d}}{\mathrm{d}y}\Phi(x) = \frac{\mathrm{d}}{\mathrm{d}x}\Phi(x) \times \frac{\mathrm{d}x}{\mathrm{d}y}$$

and

$$\psi(y) = \varphi(x) \times \frac{dx}{dy} = x\varphi(x) \ln b$$

so

$$\psi(\log_b x) = \varphi(x) \times \frac{dx}{dy} = x\varphi(x) \ln b,$$

which means that

$$\psi(\log_b ax) = ax\varphi(ax) \ln b.$$

Using the definition of scale invariance we then have

$$\begin{aligned}\psi(\log_b ax) &= ax\varphi(ax) \ln b \\ &= ax\frac{1}{a}\varphi(x) \ln b \\ &= x\varphi(x) \ln b \\ &= \psi(\log_b x).\end{aligned}$$

Therefore,

$$\psi(\log_b x + \log_b a) = \psi(\log_b x) \quad \text{and} \quad \psi(y + \log_b a) = \psi(y).$$

Since a can be chosen to be anything we wish, $\psi(y)$ repeats itself over arbitrary intervals and it can only be that it is constant. The logarithm of a scale invariant variable has a constant probability density function.

We can now relate this to the first digit phenomenon by expressing the numbers in scientific notation $x \times 10^n$, where $1 \leqslant x < 10$: the first a significant digit d of the number is simply the first digit of x. As we scale the number, we scale x, adjusting its value modulo 10. In this way, we can always think that $1 \leqslant x < 10$ whether scaled or not and if we take the base of the logarithms to be 10, $y = \log_{10} x$ will have a constant probability density function of 1 defined on $[0, 1]$. Therefore, assuming the

Table 16.4.

Digit	First digit	Second digit
0	—	0.1197
1	0.3010	0.1139
2	0.1761	0.1088
3	0.1249	0.1043
4	0.0969	0.1003
5	0.0792	0.0967
6	0.0669	0.0934
7	0.0580	0.0904
8	0.0512	0.0876
9	0.0458	0.0850

scale invariance above and for $n \in \{1, \ldots, 9\}$,

$$
\begin{aligned}
P(d = n) &= P(n \leqslant x < n + 1) \\
&= P(\log_{10} n \leqslant \log_{10} x < \log_{10}(n + 1)) \\
&= P(\log_{10} n \leqslant y < \log_{10}(n + 1)) \\
&= (\log_{10}(n + 1) - \log_{10} n)1 \\
&= \log_{10}\left(\frac{n + 1}{n}\right) \\
&= \log_{10}\left(1 + \frac{1}{n}\right),
\end{aligned}
$$

which is Benford's Law.

The reader may well detect the Weyl Equidistribution Theorem lurking here!

Extending the Law

Newcomb's general arguments in the paper mentioned earlier, which understandably he framed in terms of logarithms, led him to a table, described by the phrase

> We thus find the required probabilities of occurrence in the case of the first two significant digits of a natural number to be [as reproduced in table 16.4].

We can see that his first column heralded the Benford Law figures that we have derived: to establish the second we can proceed as follows.

If we isolate the first two significant digits of a number by writing the number as $x_1 x_2 \times 10^n$, where $10 \leqslant x_1 x_2 \leqslant 99$, and define the random variable X accordingly, we have

P(First significant digit is x_1 and the second is x_2)

$$= P(x_1 x_2 \leqslant X < x_1 x_2 + 1)$$

$$= \log_{10}\left(1 + \frac{1}{x_1 x_2}\right).$$

Now observe that

P(Second digit is x_2)

$\quad = P$(first significant digit is 1 and the second is x_2)

$\quad\quad + P$(first significant digit is 2 and the second is x_2)

$\quad\quad + \cdots$

$\quad\quad + P$(first significant digit is 9 and the second is x_2)

and we have the result

$$P(\text{Second digit is } x_2) = \sum_{r=1}^{9} \log_{10}\left(1 + \frac{1}{x_r x_2}\right).$$

A small computation yields the second column of the table and the reader may wish to pursue matters further to establish the truth of another of his statements in the paper:

> In the case of the third figure the probability will be nearly the same for each digit, and for the fourth and following ones the difference will be inappreciable.

Other results can be inferred too. For example, using that standard definition of conditional probability once again,

$$P(A \mid B) = \frac{P(A \text{ and } B)}{P(B)},$$

P(2nd s.d. is x_2 | 1st s.d. is x_1)

$$= \log_{10}\left(1 + \frac{1}{x_1 x_2}\right) \bigg/ \log_{10}\left(1 + \frac{1}{x_1}\right).$$

So, for example, the probability that the second digit of a number is 5 given that its first digit is 6 is $\log_{10}(1 + \frac{1}{65})/\log_{10}(1 + \frac{1}{6}) = 0.0990$, whereas if it started with 9 the probability is $\log_{10}(1 + \frac{1}{95})/\log_{10}(1 + \frac{1}{9}) = 0.0994$.

The most likely start to a number turns out to be 10, with a probability of $\log_{10}(1 + \frac{1}{10})/\log_{10}(1 + \frac{1}{1}) = 0.1375$.

Finally, this has been seen to be more than a theoretical curiosity. Most particularly, Mark Nigrini has pioneered its use in accounting. Quoting from him:

> Here are some possible practical applications for Benford's Law and digital analysis.
>
> - Accounts payable data.
> - Estimations in the general ledger.
> - The relative size of inventory unit prices among locations.
> - Duplicate payments.
> - Computer system conversion (for example, old to new system; accounts receivable files).
> - Processing inefficiencies due to high quantity/low dollar transactions.
> - New combinations of selling prices.
> - Customer refunds.

Which is rather more than Newcomb envisaged, to judge by this final quotation from his paper:

> It is curious to remark that this law would enable us to decide whether a large collection of independent numerical results were composed of natural numbers or logarithms.

Chapter 17

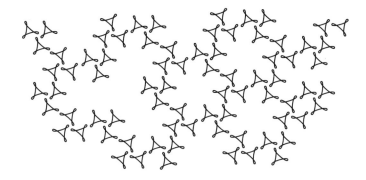

GOODSTEIN SEQUENCES

Bigger than the biggest thing ever and then some. Much bigger than that in fact, really amazingly immense, a totally stunning size, real 'wow, that's big', time... Gigantic multiplied by colossal multiplied by staggeringly huge is the sort of concept we're trying to get across here.

Douglas Adams, *The Restaurant at the End of the Universe*

Exponential notation deals with very big numbers very efficiently. For example, $2^{2^{2^2}}$ has about one million digits in it (and happens to be the biggest number which can be manufactured from four 2s using the standard arithmetic operations). In Adams's original book of the *Hitchhiker* series, *The Hitchhiker's Guide to the Galaxy*, appears what might be the biggest number ever used in a work of fiction: $2^{260\,199}$, the quoted odds against Arthur Dent and Ford Prefect being rescued by a passing spaceship, having been thrown out of an airlock. (In fact, they were rescued by a spaceship—powered by an 'infinite improbability drive'.)

Exponential notation can be thought to be the third in the progression of the basic arithmetical operations, as addition is built on by multiplication and that in turn by exponentiation: $(2 \times 5 = 2 + 2 + 2 + 2 + 2)$ and $(2^5 = 2 \times 2 \times 2 \times 2 \times 2)$. It seems natural to extend the process to a fourth operation by defining repeated exponentiation: that operation is commonly given the name *tetration* (*tetra-* the Greek for four together with a part of the word *iteration*). A notation that has been commonly used is to put the power to the left of the base number, so, for example,

$$^{3}2 = 2^{(2^2)} = 2^4 = 16,$$

$$^{4}2 = 2^{(2^{(2^2)})} = 2^{16} = 65\,536,$$

$$^{5}2 = 2^{(2^{(2^{(2^2)})})} = 2^{65\,536} = \text{a very large number indeed.}$$

Notice that these 'power towers' are evaluated unambiguously from the highest power down.

Alternative names and notations exist which replace and extend the above, for example, Knuth's *up-arrow* and Conway's *chained-arrow* notations, but it was the English mathematical logician, philosopher and teacher Reuben Goodstein who coined the term tetration and it was he who discovered a most remarkable fact involving fantastically large numbers—and also a proof which is equally remarkable.

Goodstein Sequences

We count in base 10. This means that, for example, the number 2136 has the natural decomposition

$$2136 = 2 \times 10^3 + 1 \times 10^2 + 3 \times 10^1 + 6 \times 10^0$$

and, generally, that $(a_1 a_2 a_3 \cdots a_n)_{10} = \sum_{r=1}^{n} a_r \times 10^{n-r}$ for positive integers $a_1, a_2, a_3, \ldots, a_n < 10$.

Evidently, we could use any positive integer for the base: with base 2 we have

$$2136 = 2^{11} + 2^6 + 2^4 + 2^3,$$

where the nonzero a_r are necessarily 1. We can 'complete' this binary representation by writing the powers themselves in binary to arrive at a first stage of

$$2136 = 2^{2^3+2+1} + 2^{2^2+2} + 2^{2^2} + 2^{2+1}$$

and a second of

$$2136 = 2^{2^{2+1}+2+1} + 2^{2^2+2} + 2^{2^2} + 2^{2+1},$$

which is known as the *complete base 2* representation of 2136.

Now we move to the more esoteric and define a set of 'base bumping' functions, starting with B_3, which acts on the complete base 2 representation of 2136 by replacing each 2 with a 3, to give

$$B_3(2136) = 3^{3^{3+1}+3+1} + 3^{3^3+3} + 3^{3^3} + 3^{3+1} \approx 3.6 \times 10^{40},$$

a number vastly bigger than the original.

From this we will define the Goodstein sequence $G_r(n)$ by

$$G_r(n) = \begin{cases} n \text{ written in complete base 2,} & r = 2, \\ B_r(n) - 1 \text{ (simplified),} & r > 2, \end{cases}$$

which means that we subtract 1 from our number to get

$$G_3(2136) = 3^{3^{3+1}+3+1} + 3^{3^3+3} + 3^{3^3} + 3^{3+1} - 1,$$

and to maintain the proper base representation we have to re-arrange the 3^{3+1} term a bit, a task achieved by noting that from the theory of geometric series, for any positive integer b,

$$b^n - 1 = (b-1) \sum_{r=1}^{n} b^{r-1} = (b-1) \sum_{r=1}^{n} b^{n-r} = \sum_{r=1}^{n} (b-1)b^{n-r}.$$

This makes

$$3^{3+1} - 1 = 2 \times 3^3 + 2 \times 3^2 + 2 \times 3 + 2$$

and

$$G_3(2136) = 3^{3^{3+1}+3+1} + 3^{3^3+3} + 3^{3^3} + 2 \times 3^3 + 2 \times 3^2 + 2 \times 3 + 2.$$

We continue the process with 2136 to get

$$G_4(2136) = B_4(2136) - 1$$
$$= 4^{4^{4+1}+4+1} + 4^{4^4+4} + 4^{4^4} + 2 \times 4^3 + 2 \times 4^2 + 2 \times 4 + 1$$
$$\approx 3.3 \times 10^{619}$$

and

$$G_5(2136) = B_5(2136) - 1$$
$$= 5^{5^{5+1}+5+1} + 5^{5^5+5} + 5^{5^5} + 2 \times 5^3 + 2 \times 5^2 + 2 \times 5$$
$$\approx 4.0 \times 10^{10\,925}.$$

It seems evident that, as the iteration continues, the resulting Goodstein number will become greater and greater, but let us see what happens as we start with the smallest positive integers.

First, $G_2(1) = 1$ and so $G_3(1) = 1 - 1 = 0$ and we are finished. Now consider the Goodstein sequence generated by 2:

$$G_2(2) = 2^1, \quad G_3(2) = 3^1 - 1 = 2, \quad G_4(2) = 2 - 1 = 1$$

and finally $G_5(2) = 1 - 1 = 0$.

Again, the sequence collapses to 0. We have to work a little harder with 3: $G_2(3) = 2^1 + 1$, $G_3(3) = 3^1$, $G_4(3) = 4^1 - 1 = 3$, $G_5(3) = 2$, $G_6(3) = 1$ and finally $G_7(3) = 0$ and the collapse occurs once more.

A Rather Big Surprise

How big must the integer be before these fantastically large numbers appear? Let us see what happens with 4. $G_2(4) = 2^2 = 4$, $G_3(4) = 3^3 - 1 = 2 \times 3^2 + 2 \times 3 + 2 = 26$, $G_4(4) = 2 \times 4^2 + 2 \times 4 + 1 = 41$ and continuing the procedure leads to the extended sequence:

$$r = \quad 2 \quad 3 \quad 4 \quad 5 \quad 6 \quad 7 \quad 8 \quad \cdots,$$
$$G_r(4) = \quad 4 \quad 26 \quad 41 \quad 60 \quad 83 \quad 109 \quad 139 \quad \cdots.$$

The values of the Goodstein sequence are now growing, albeit slowly, and we have a start—which continues in a manner rather stranger than one might imagine. After precisely $3 \times 2^{402\,653\,211} - 1 \approx 10^{121\,210\,695}$ steps the sequence once again collapses to 0.

That is, if $r = 3 \times 2^{402\,653\,211} - 1$, $G_r(4) = 0$, vastly big though the terms become!

David Williams has provided an argument to establish this remarkable fact, based on observations of the patterns in the Goodstein sequence of the number 4. The reader may wish to check his observations that:

1. For $r \leqslant 27$, if $m = 3 \times 2^r - 1$, then $G_m(4) = 1 \times m^2 + (27 - r) \times m$.

2. This means that, if $a = 3 \times 2^{27} - 1$, then $G_a(4) = a^2$.

3. Write $b = a + 1 = 3 \times 2^{27}$ and this result becomes $G_b(4) = b^2 - 1 = (b - 1)b + (b - 1)$.

4. For $r \leqslant b - 1$, this pattern continues to $G_{b+r}(4) = (b - 1)(b + r) + b - (r - 1)$.

5. This means that $G_{2b-1}(4) = (b - 1)(2b - 1)$.

6. If $g = 2^s b - 1$, this pattern continues to $G_g(4) = (b - s)g$.

7. This means that, when $s = b$, $G_g(4) = 0$.

8. Which means that $G_g(4) = 0$ for the first time when

$$g = 2^b b - 1 = 2^{3 \times 2^{27}} \times (3 \times 2^{27}) - 1 = 3 \times 2^{402\,653\,211} - 1.$$

Now that we have the Goodstein sequence for 4 (eventually) reaching 0 we will make the vast jump to the earlier sequence generated by 2136. Is it possible that this would eventually finish at 0 too, huge though the numbers become? Astonishingly, the answer is *yes*, in fact every such sequence would do the same. This 1944 result of Reuben Goodstein (On the restricted ordinal theorem, *Journal of Symbolic Logic* 9:33–41) is, understandably, known as *Goodstein's Theorem* and simply states that *every Goodstein sequence converges to* 0.

'Bumping the base' seems inevitably to increase the size of the number hugely and subtracting 1 hardly seems to compensate for this—but the process is deceptive!

Axioms and Ordinals

The result is included here because of its surprising nature, but its importance to mathematical logic is of vastly greater moment since it provides an example of what have become known as *naturally independent phenomena*, which have come about as a by-product of the seminal work of Kurt Gödel. At a conference in Königsberg in September 1930, he announced his first *incompleteness theorem*, which destroyed the long-cherished notion of the great David Hilbert that mathematics was complete within itself; that is, that every statement expressible within it can be proved or disproved within it. Yet there remained a degree of dissatisfaction. Gödel's construction of an undecidable statement was 'meta-mathematical' and there was felt to be a need for an example of such a statement which was not peculiar or contrived, one which occurs 'naturally' and which we would reasonably expect to be decidable one way or the other: that need was fulfilled by versions of *Ramsey's Theorem*, *Kruskal's Tree Theorem* and *Goodstein's Theorem*, which had to wait until 1982 for J. Paris and L. Kirby to show that it is not provable within ordinary arithmetic. Since Goodstein had indeed proved his theorem back in 1944 we seem to have irreconcilable statements:

- It *is* true, I just proved it (Goldstein 1944).
- It is not provable one way or the other (Paris and Kirby 1982).

Of course, the reconciliation exists and it is found in a clear definition of what we mean by *mathematics* in this context and the somewhat exotic nature of Goodstein's proof: the use of transfinite ordinals. The 'ordinary arithmetic' involved in Paris and Kirby's argument is formally defined by the Peano axioms, which include all of the familiar rules of algebra together with the principle of induction; it is all that we need to cope with everyday mathematics. Set theory is not part of it and, most particularly, infinite sets are not accounted for; there is no place in it for Cantor's transfinite ordinals. We saw some of the wonder of Cantor's work in chapter 6, now we will very briefly look at some more products of his original mathematical mind.

His construction of transfinite numbers distinguishes between the use of the positive integers as a measure of size (for example, there are five elements in the set) and order (for example, that is the fifth element of the set) and here we are concerned with this latter interpretation.

Set theory (using the empty set $\emptyset$, a letter from the Norwegian alphabet apparently chosen by Andre Weil as part of the remarkable Bourbaki initiative) can be used to define the finite ordinals in the following recursive way:

$$0 \equiv \emptyset, \quad 1 \equiv \{\emptyset\} \equiv \{0\}, \quad 2 \equiv \{\emptyset, \{\emptyset\}\} \equiv \{0, 1\},$$
$$3 \equiv \{\emptyset, \{\emptyset\}, \{\emptyset, \{\emptyset\}\}\} \equiv \{0, 1, 2\}, \quad \text{etc.,}$$

and this suggests the ordering $a \leqslant b$ if and only if $a \subseteq b$ and with this we retrieve the familiar, natural ordering $0, 1, 2, 3, 4, \ldots$.

Things become a little more interesting when we consider the whole set of natural numbers $\{0, 1, 2, 3, \ldots\}$ and define the first transfinite ordinal $\omega = \{0, 1, 2, 3, \ldots\}$ and the extended ordinal sequence $0, 1, 2, 3, 4, \ldots, \omega$. This ω is distinguished in that it has no immediate predecessor and it is clear from the definition that $n < \omega$ for all integers n. The process continues as we define $\omega + 1 \equiv \{0, 1, 2, 3, \ldots, \omega\}$, etc., to achieve the sequence

$$\{0, 1, 2, 3, 4, \ldots, \omega, \omega + 1, \omega + 2, \omega + 3, \ldots\},$$

which itself continues to

$$\{0, 1, 2, 3, 4, \ldots, \omega, \omega + 1, \omega + 2, \omega + 3, \ldots, \omega + \omega \equiv \omega \times 2\}$$

and thence to

$$\{0, 1, 2, 3, 4, \ldots, \omega, \omega + 1, \omega + 2, \omega + 3, \ldots,$$
$$\omega + \omega = \omega \times 2, \omega \times 2 + 1, \ldots, \omega \times 2 + \omega = \omega \times 3, \ldots,$$
$$\omega \times 4, \ldots, \omega \times \omega = \omega^2, \omega^2 + 1, \ldots, \omega^2 + \omega, \ldots,$$
$$\omega^2 + \omega \times 2, \ldots, \omega^3, \ldots, \omega^4, \ldots, \omega^\omega, \ldots\}$$

with the sequence continuing to (and beyond) the power-tower limit $\varepsilon_0 = \omega^{\omega^{\omega^{\cdots}}}$, which is designated 'epsilon zero'. The familiar notation used to represent this ordinal sequence is deceptively subtle; for example, $1 + \omega = \omega \neq \omega + 1$ and $2 \times \omega = \omega \neq$

$\omega \times 2$, but we sidestep this important matter (it depends on what is meant by '=') to concentrate on a crucially important property of the set of transfinite ordinals: with the ordering above they are *well ordered* by $\leqslant$, a condition defined on all ordinals a, b and c by

1. $a \leqslant a$.
2. If $a \leqslant b$ and $b \leqslant c$, then $a \leqslant c$.
3. If $a \leqslant b$ and $b \leqslant a$, then $a = b$.
4. Either $a \leqslant b$ or $b \leqslant a$.
5. Every nonempty subset of the ordinals has a least element.

These seemingly innocent conditions on a set, and in particular on the set of ordinals, conceal an important consequence: in such a set there can be no infinite descending chains, that is, $a \geqslant b \geqslant c \geqslant \cdots$ must be of finite length.

Goodstein's Argument

With the Goodstein process acting on a positive integer, we have a procedure which is plainly number theoretic. The assertion that the sequence inevitably converges to 0 seems to be one which can be resolved using the many powerful results of number theory, but Paris and Kirby proved otherwise: to prove the result we have to move to these transfinite ordinals, and that is what Goodstein did. The following argument puts all of the pieces in place.

First, every transfinite ordinal $a < \varepsilon_0$ can be written in a manner rather like the complete base representation of integers, using base ω; for example, $\omega^{\omega^\omega} + \omega^{\omega+1} + \omega$. This is called *Cantor's normal form* of the ordinal.

Now we define the sequence $G_r^\omega(n)$ parallel to the Goodstein sequence $G_r(n)$ to be the sequence of transfinite ordinals generated by replacing the base of the Goodstein number with ω.

For example, since $G_2(2136) = 2^{2^{2+1}+2+1} + 2^{2^2+2} + 2^{2^2} + 2^{2+1}$,

$$G_2^\omega(2136) = \omega^{\omega^{\omega+1}+\omega+1} + \omega^{\omega^\omega+\omega} + \omega^{\omega^\omega} + \omega^{\omega+1}.$$

Now let us begin to get a feel for this new sequence by looking at what form it takes with the number 4. The previous results were that $G_2(4) = 2^2$, $G_3(4) = 2 \times 3^2 + 2 \times 3 + 2$, $G_4(4) = 2 \times 4^2 + 2 \times 4 + 1, \ldots$ and this means that

$$G_2^\omega(4) = \omega^\omega, \quad G_3^\omega(4) = 2 \times \omega^2 + 2 \times \omega + 2,$$
$$G_4^\omega(4) = 2 \times \omega^2 + 2 \times \omega + 1.$$

Notice that this sequence of ordinals is descending:

$$\omega^\omega > 2 \times \omega^2 + 2 \times \omega + 2 > 2 \times \omega^2 + 2 \times \omega + 1 \cdots.$$

If the Goodstein sequence were of infinite length, it would generate this parallel sequence of descending ordinals which is infinite in length... and that contradicts the well-ordering result stated on page 207. Of course, this means that the Goodstein sequence of 4 must terminate in 0.

It is not very difficult to generalize the process to any Goodstein sequence

$$G_2(n), \ G_3(n), \ G_4(n), \ G_5(n), \ \ldots, \ G_k(n), \ \ldots$$

being paralleled by the decreasing ordinal sequence

$$G_2^\omega(n), \ G_3^\omega(n), \ G_4^\omega(n), \ G_5^\omega(n), \ \ldots, \ G_k^\omega(n), \ \ldots$$

to provide proof of the theorem.

The result provides a striking realization of Cantor's own view that 'the essence of mathematics lies in its freedom'.

Chapter 18

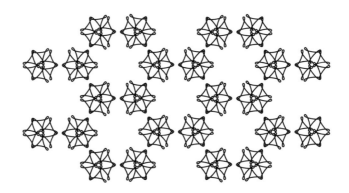

THE BANACH-TARSKI PARADOX

God exists since mathematics is consistent, and the Devil
exists since we cannot prove it.

Andre Weil

The subject matter of this last chapter simply has to rank as
the most counterintuitive result in mathematics and is a fitting
finalé to a book devoted to mathematical surprise.

Stefan Banach and Alfred Tarski brought to the world an
improvement on a paradox devised by the great topologist Felix
Hausdorff, the formalized form of which is often replaced by
something fanciful such as:

A solid sphere can be dissected into five pieces and the
pieces reassembled to form two complete spheres of exactly
the same size as the original.

Or, alternatively:

A solid sphere the size of a pea can be dissected into finitely many pieces which can be rearranged using rotations and translations to form a solid sphere the size of the Sun.

To those who have not seen the result(s) before, it must seem as if there is a misprint—or that the author has allowed himself to become a little too carried away. Actually, what has been written above *is* true and we will attempt to give the reader some flavour of why it is so.

Formalization

We must formalize things, and to that end we define the Euclidean '3 ball' of radius r, B_r, by $B_r = \{x \in \mathbb{R}^3 : |x| \leqslant r\}$ and agree that a *rigid motion* of $\mathbb{R}^3$ is a transformation R which preserves Euclidean distance (i.e. is such that for all points $x, y \in \mathbb{R}^3$, $|x - y| = |R(x) - R(y)|$. Now we can state the results more formally as:

> There exists a decomposition of B_r into five pairwise disjoint sets A_1, A_2, A_3, A_4, A_5 (of which the last is a single point) such that there exist rigid motions R_1, R_2, R_3, R_4, R_5 with $B_r = R_1(A_1) \cup R_2(A_2)$ and $B_r = R_3(A_3) \cup R_4(A_4) \cup R_5(A_5)$, where all unions are disjoint.

Or:

> For any two distinct positive integers m and n, B_m can be split into a disjoint union of sets $A_1, \ldots, A_n$ such that there exist rigid motions $R_1, \ldots, R_n$ so that $B_n = R_1(A_1) \cup R_2(A_2) \cup \cdots \cup R_n(A_n)$, where all the unions are disjoint.

There is also its most general form in $\mathbb{R}^3$:

> Any two bounded subsets of $\mathbb{R}^3$ (with nonempty interior) can be dissected and reassembled each to form the other.

Given that these three statements are true (and indeed they are), one must expect a catch (otherwise it would not have taken Jesus Christ to supposedly feed the five thousand with five loaves and eight fishes): the catch is that in this case the proof is non-constructive, it demonstrates existence without revealing how to achieve the aim. It also demonstrates that this is intrinsically a mathematical result and it cannot actually be realized. Put another way, in the decomposition, the pieces will not be *measurable* and so they will not have *reasonable* boundaries or a well-defined *volume* in the accepted sense of the term. More plainly, it is impossible to carry out the dissection since cutting with a knife creates only measurable sets.

The Axiom of Choice

At the heart of the proof of the result lies the *axiom of choice.* This most deceptive statement, formulated about a century ago by the mathematical logician Ernst Zermelo, simply states that

> In any collection of nonempty sets, we can form a new set
> by *choosing* a member from each set in that collection.

It seems very obvious that such a thing is possible, but everything depends on what we mean by the word *choosing.* For example, for any finite collection of nonempty sets we can form a new set by choosing the first element of each of the sets. Moving to infinite sets could be more problematic, but consider the collection of all nonempty subsets of the natural numbers $\{0, 1, 2, 3, \ldots\}$, then we can form our new set by choosing the smallest element in each set. A little more subtly, consider the collection of all nonempty subintervals of $(0, 1)$, there remains no problem since we can form our new set by choosing the midpoint of each interval. So where *is* the problem? Actually, so far there isn't one, since the axiom of choice has not come into play; in each case we had a rule for doing the choosing. Now, for example, take the collection of all nonempty subsets of $\mathbb{R}$, in which case there is no consistent procedure for choosing the elements and thereby populating our set from the infinite number of subsets. Now we need that guarantee, provided by the axiom of choice, which simply states that there is some procedure which allows us to *choose* an element from each set in the collection—and never mind what the procedure is; it gives no indication of how the choosing would be done; it simply guarantees the existence of that choice. Notice also that it is called an *axiom*; that is, it is an assumption. It has been shown that, if the standard axioms of set theory are consistent without the axiom of choice, they remain so if it is included. This means that we can have one system of mathematical logic in which the axiom features and a completely different one in which it does not (rather like the geometries arising for the parallel postulate, mentioned on page 70).

The axiom has many equivalent forms, some (subjectively) more 'obviously true' than others, possibly the most important of which are the *well-ordering principle* and the considerably more enigmatic *Zorn's Lemma*:

> *The well-ordering principle.* A set is said to be *well ordered* if every subset of it has a first element. The well-ordering principle states that every set can be well ordered (which we came across in the previous chapter).

> *Zorn's Lemma.* Every *partially ordered* set in which every *totally ordered* subset has an upper bound contains at least one maximal element.

We will leave the interested reader to dissect this, but this is enough to appreciate mathematician Jerry Bona's quip about these entirely equivalent statements:

> The axiom of choice is obviously true, the well-ordering principle obviously false, and who can tell about Zorn's Lemma?

Groups

The axiom of choice is not alone sitting at the heart of the proof of the paradox; that proof also relies fundamentally on the concept of a *group*, and in particular a *free group on two generators* and also that of a *rotation group*—and here is where problems of exposition begin to become overwhelming. It is impossible in such a tiny space to give anything like a representative overview of the fundamental algebraic abstraction of a group, the minimal system required for useful abstract algebra to exist. The definition is itself minimal though, but it efficiently conceals the great significance of the idea: a group G is any set of objects, together with a law of combination ($*$), which satisfy:

1. For all $a, b \in G$, $a * b \in G$ (closure).
2. For all $a \in G$, there exists an element $\varepsilon \in G$ so that $a * \varepsilon = \varepsilon * a = a$ (identity element).
3. For each $a \in G$, there exists an element $a^{-1} \in G$ so that $a * a^{-1} = a^{-1} * a = \varepsilon$ (inverse elements).
4. For all $a, b, c \in G$, $a * (b * c) = (a * b) * c$ (associativity).

As the prototype case just consider the set of all integers with $*$ replaced by $+$, ε by 0 and a^{-1} by $-a$.

Notice that the definition does not include the assumption of commutivity, that is, it is not assumed that $a * b = b * a$, although this might be the case.

A particular case of a group is one which is formed by formally combining any number of abstract symbols (generators), with or without a law or laws which specify how combinations of elements should be simplified; if no such law exists such a group has the name of a *free group* on however many generators there might be. For example, the free group on two generators a and b consists of all finite strings that can be formed from the five symbols ε, a, a^{-1}, b, b^{-1} (inverses have to be included) such that no a appears directly next to an a^{-1} and no b appears directly next to a b^{-1} (since these must both simplify to ε). Two such strings can be concatenated and simplified to a string of this type by repeatedly making use of the cancellation brought about by combining an element with its inverse wherever possible. For instance, $aba^{-1}b^{-1}a$ concatenated on the right with $a^{-1}ba^{-1}b^{-1}a$ results in

$$aba^{-1}b^{-1}aa^{-1}ba^{-1}b^{-1}a = aba^{-1}a^{-1}b^{-1}a.$$

We will represent this group by the letter G; it is such a group that is required for the Banach–Tarski paradox to be established. (For the sake of contrast, a condition that might be placed on a group on two generators might be a way of rewriting the element ba, for example, $ba = a^2b$.)

The Paradox

The group G can be 'paradoxically decomposed' as follows. Let $S(a)$ be the set of all strings that start with a and define $S(a^{-1})$, $S(b)$ and $S(b^{-1})$ similarly. Since every element of G must either be ε or start with one of these four symbols, it must be the case that

$$G = \{\varepsilon\} \cup S(a) \cup S(a^{-1}) \cup S(b) \cup S(b^{-1}),$$

but notice that G can also be divided into $S(a)$ and the rest of the elements—and that these are all elements which do not begin with a and which can therefore be written as $aS(a^{-1})$; this means that G can also be written in the form $G = aS(a^{-1}) \cup S(a)$ and for the same reason $G = bS(b^{-1}) \cup S(b)$. This seemingly simple observation will bring about the paradox.

Now we realize G as a group of rotations of $\mathbb{R}^3$ by choosing two perpendicular axes and defining element a to be a rotation of $\cos^{-1}\frac{1}{3}$ about the first and b to be a rotation of $\cos^{-1}\frac{1}{3}$ about the second—a step which cannot be performed in two dimensions. It is not obvious but these do form our free group on two generators and the paradoxical decomposition above applies to this form of G.

Now apply G to the sphere $S_r = \{x \in \mathbb{R}^3 : |x| = r\}$ by taking points on it and rotating them accordingly and collect together all points x_1, x_2 which are such that $x_1 = gx_2$ for some $g \in G$; that is, we partition the sphere into *orbits* brought about by the action of G, with two points belonging to the same orbit if and only if there is a rotation in G which moves the first point into the second. Now we need that axiom of choice. Use it to pick exactly one point from every orbit and let these points form a set X. It is the case that almost every point in S_r can be reached in exactly one way by applying the proper rotation from G to the proper element from X and, because of this, the paradoxical decomposition of G then yields a paradoxical decomposition of S_r.

Finally, connect every point on S_r with a ray to the origin and so generate an infinite number of spheres; the paradoxical decomposition of S_r then yields a paradoxical decomposition of the solid unit ball—minus the origin, but that is where ε comes in.

And there is the famous Banach-Tarski paradox 'proved'! There may not be any holes left in the decomposition of the sphere, but there are a number in the details of the above, but they are matters of detail and can be patched up to a complete and rigorous proof.

It is interesting to note that the proof depends on three dimensions (with that group of rotations); while intuitively the two-dimensional case seems to be easier, it is in fact not true that all bounded subsets of the plane with nonempty interior are capable of being dissected one to the other. There is one in particular which does exist though: a circular disc can be cut into finitely many pieces and reassembled to form a square of equal area—the 'circle-squares' were in some sense correct after all. The challenge to establish a way of doing this was posed by Tarski in 1925 and it took until 1990 to answer it, when Miklos Laczkovich proved it possible—using about 10^{50} different pieces.

So, the result is not practical, but the late Ralph P. Boas Jr found a use for it in his amusing and eclectic book (published in 1996 by Mathematical Association of America) *Lion Hunting and Other Mathematical Pursuits. A Collection of Mathematics, Verse and Stories*, in which over thirty different 'proven' methods are given to capture a lion. (The book was an expansion of the famous spoof paper 'A contribution to the mathematical theory of big game hunting' by one H. Petard, 1938, *American Mathematical Monthly* 45:446-47.) The one relevant to us is his idea to apply the Banach–Tarski decomposition to the lion, put the pieces back together to form a feline the size of a domesticated cat (from which we may expect only minor harm). Then hunt it fearlessly, capture it and after caging the beast, use the Banach-Tarski decomposition once again to rearrange the pieces into their original configuration!

If nothing else in this book was considered *Impossible* by the reader, it is hoped that this result might just have saved the author's day.

The Motifs

For some minutes Alice stood without speaking, looking out
in all directions over the country—and a most curious country it was. There were a number of tiny little brooks run-
ning straight across it from side to side, and the ground
between was divided up into squares by a number of little
green hedges, that reached from brook to brook.

'I declare it's marked out just like a large chessboard!'
Alice said at last.

Through the Looking-Glass, and What Alice Found There
Lewis Carroll

The design of the chapter motifs did not demand Carroll's fantastic, mutually perpendicular brooks and hedges but a pair of
infinite invisible lines, intersecting at an arbitrary angle, and
repeated at regular intervals. By this means the plane is divided
into an infinite number of congruent parallelograms, the vertices of which form an infinite, regular lattice. The two lines
determine two independent directions of translation and the
lengths of the sides of the parallelograms the fundamental translation distances, which must be bounded below by a number
$\varepsilon > 0$. It is with these two independent translations that the
study of the wallpaper groups begins; informally, we may think
of them as determining the possible ways of generating infinite
two-dimensional patterns using those translations and the other
isometries of the plane: rotations, reflections and glide reflections. The lattice points are obvious choices for centres of rotation and the lines for mirror lines, but there are other possibilities too and using them brings about the remarkable fact
that there are just seventeen essentially different patterns possible: no matter how different one wallpaper design looks from
another, if it is regular it must be of one of these types. Chapter 1

Chapter	Full HM notation	Short HM notation	Brief description
1	Basic pattern		
2	p1	p1	Consists only of translations.
3	p211	p2	Contains 180° rotations.
4	p1m1	pm	Contains reflections. One mirror line is parallel to one translation direction and the other mirror line is perpendicular to the other translation direction.
5	p1g1	pg	Contains glide reflections. The direction of the glide reflections is parallel to one translation direction and perpendicular to the other translation direction.
6	p2mm	pmm	Contains reflections with perpendicular mirror lines.
7	p2mg	pmg	Contains both a reflection and a rotation of order 2. The centres of rotations do not lie on the mirror lines.
8	p2gg	pgg	Contains glide reflections and half-turns. The glide reflections have perpendicular axes and the centres of the half-turns do not lie on these axes.
9	c1m1	cm	Contains reflections and glide reflections with parallel axes. The mirror lines bisect the angle formed by the translation directions.
10	c2mm	cmm	Contains reflection with perpendicular axes and also rotations of order 2. The centres of the rotations do not lie on the mirror lines.
11	p4	p4	Contains a rotation of orders 2 and 4. The centres of the order 2 rotations are midway between the centres of the order 4 rotations.
12	p4mm	p4m	As p4 but it also contains reflections. The mirror lines are inclined at 45° to each other so that four mirror lines pass through the centres of the order 4 rotations.

Chapter	Full HM notation	Short HM notation	Brief description
13	p4gm	p4g	Contains reflections and rotations of orders 2 and 4. The mirror lines are perpendicular, and none of the rotation centres lie on these mirror lines.
14	p3	p3	Contains a 120° rotation.
15	p3m1	p3m1	Contains reflections and rotations of order 3. The mirror lines are inclined at 60° to one another, and all of the centres of rotation lie on the mirror lines.
16	p31m	p31m	Contains reflections whose axes are inclined at 60° to one another and rotations of order 3. Some of the centres of rotation lie on the mirror lines and some do not.
17	p6	p6	Contains rotations of orders 2, 3 and 6.
18	p6mm	p6m	Contains rotations of order 2, 3 and 6 as well as reflections. The mirror lines meet at all the centres of rotation. At the centres of the order 6 rotations, six mirror lines meet and are inclined at 30° to one another.

begins with a basic design, a simple figure based on an equilateral triangle; the following seventeen chapters each begin with a design generated from the figure by the action of one of the seventeen possible wallpaper transformations.

It seems to have been the French eclectic Camille Jordan who first catalogued wallpaper groups in 1869, through his investigations into crystallographic structure: he identified sixteen of the seventeen possibilities: the full set was listed by the Russian crystallographer E. S. Federov in 1891.

The seemingly innocent ε condition on the translations is called the *discreteness condition*, which ensures that the fundamental parallelogram has a well-defined, finite area. This and the independence of the two translations carries another important and nontrivial consequence: the *crystallographic restriction theorem*, which states that, if there are rotations in a wallpaper pattern, they must be of order 2, 3, 4 or 6; that is, every rotation must be by 180°, 120°, 90° or 60°.

A representative discussion of the matter would occupy too much space here but we encourage the interested reader to pursue the matter further, as there are plenty of sources available, for example, the classic books *Geometry and the Imagination* by David Hilbert and Stephan Cohn-Vossen and *Introduction to Geometry* by H. S. M. Coxeter deal with the matter superbly and, for those who would enjoy a completely rigorous if abstract approach, then the article by R. L. E. Schwarzenberger (The 17 plane symmetry groups, 1974, *The Mathematical Gazette* 58(404):123–31) might well suit.

The different patterns are coded by what is called *Hermann–Mauguin* (HM) notation, named after the German crystallographer Carl Hermann and the French mineralogist Charles-Victor Mauguin. We will content ourselves with the table on pages 218 and 219, which identifies the patterns chapter by chapter.

Appendix

Principle of Induction

In Singapore, on Monday 18 August 2003, the South Korean ambassador to Singapore knocked over the first of more than 303 000 dominos to create a new world record for the longest solo dominoes topple. A Chinese lady, Ma Li Hua, had created the arrangement, having spent six weeks, working twelve hours a day doing so; they took just six minutes to fall.

The success of the attempt required two things: Ms Li Hua's prodigious industry in ensuring that, if one domino should topple, the one following would topple too and the ambassador's contribution in toppling the first domino. With these two actions assured, so was the success of the endeavour. There is a mathematical equivalent, a jewel in the crown of mathematical techniques: induction, which this section will attempt briefly to explain. In its most basic form, the principle of induction states:

> Suppose that a proposition $P(n)$ is defined for all positive integers n (or some ordered infinite subset of them) and that we know $P(1)$ to be true. If we assume that $P(r)$ is true (for some $r > 1$) and can conclude that $P(r+1)$ is therefore true, then $P(n)$ is true for all integers.

To prove the result in each case we need to 'knock over the first domino' by showing that $P(n)$ is true for $n = 1$ and, having done that, ensure that if one domino falls, so does the next one; that is, if $P(n)$ is true for $n = k$ it must be true for $n = k + 1$.

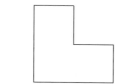

Figure 1.

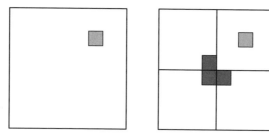

Figure 2.

Although there is evidence that the principle was understood several centuries before, the first accepted proof by mathematical induction appears in 1575 in Francesco Maurolico's *Arithmeticorum Libri Duo*, where he established that the sum of the first n odd integers is n^2.

Here $P(n)$ is the proposition

$$1 + 3 + 5 + \cdots + (2n - 1) = n^2.$$

Since $1 = 1^2$, $P(1)$ is true and we have 'knocked over the first domino'.

Now suppose that the kth domino falls, that is, $P(k)$ is true and therefore that

$$1 + 3 + 5 + \cdots + (2k - 1) = k^2$$

and consider

$$[1 + 3 + 5 + \cdots + (2k - 1)] + (2k + 1) = k^2 + (2k + 1) = (k + 1)^2$$

and this means that the $(k + 1)$th domino falls and so $P(k + 1)$ is true: all dominos fall and the statement is true for all natural numbers.

To gain something of an appreciation for the diversity of the technique, we will look at two further examples of its use for statements defined on the natural numbers:

(i) $P(n)$ is the statement, 'If $y = x^n$, $dy/dx = nx^{n-1}$' for n a positive integer.

Since

$$y = x^1 \Rightarrow \frac{dy}{dx} = 1 = 1x^0,$$

we have $P(1)$ is true. Now suppose that $P(k)$ is true, that is,

$$y = x^k \Rightarrow \frac{dy}{dx} = kx^{k-1}.$$

Now consider $y = x^{k+1} = x \times x^k$. Using the product rule and our inductive assumption,

$$\frac{dy}{dx} = 1 \times x^k + x \times kx^{k-1} = x^k + kx^k = (k+1)x^k,$$

which makes $P(k+1)$ true.

(ii) $P(n)$ is the statement, 'A $2^n \times 2^n$ chessboard with any one square deleted can be covered with 2×1 L-shaped pieces as shown in figure 1'.

Here, $P(1)$ is the proposition that a 2×2 chessboard with any square missing can be covered with such a 2×1 L-shaped piece; this is obvious, as the chessboard becomes the piece. Now suppose that $P(k)$ is true, that is, a $2^k \times 2^k$ chessboard with any one square deleted can be covered with 2×1 L-shaped pieces and consider a $2^{k+1} \times 2^{k+1}$ chessboard with a square missing. Now divide this chessboard into four equal quadrants, as shown in figure 2, each of size $2^k \times 2^k$, and place one of the L-shaped pieces at the centre so that it lies within the three quadrants not containing the empty square. By the inductive hypothesis, it is possible to tile each of the four quadrants: the one with the square missing and the three with a square effectively missing since it has already been covered. This means that the induction is complete, with $P(k+1)$ true.

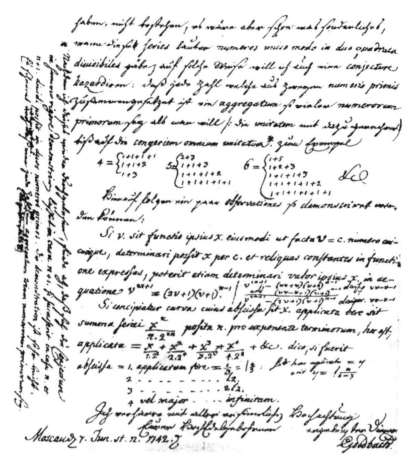

Figure 3.

The Goldbach Conjecture

On 7 June 1742 Leonard Euler, one of the greatest mathematicians of all time—and arguably the most prolific—received a letter, shown in figure 3, from one Christian Goldbach, a Prussian amateur mathematician and historian, with whom Euler had a regular correspondence.

The margin of the letter contains the sentence:

Es scheinet wenigstens, daß eine jede Zahl, die größer ist als 2, ein aggregatum trium numerorum primorum sey.

This translates as 'the conjecture that every number greater than 2 is the sum of three primes.' For this to be sensible, it must have been the case that Goldbach considered 1 to be prime, a convention that is no longer followed (since unique factorization of integers is greatly important). Its most common, modern equivalent form is that every even positive integer greater than or equal to 4 can be written as the sum of two primes. At the time of this book going to print it remains a conjecture, although there has been at least one 'proof'. However, this has not met with general acceptance (the publisher Faber has offered $1,000,000 for an accepted proof). The strongest result to date is one of L. G. Schnirelman, when in 1939 he proved that every even positive integer can be written as the sum of not more than 300 000 primes! It is known that the conjecture holds for integers less than 10^{18} and that will be plenty big enough for the purposes to which we put the conjecture.

The Exponential and Trigonometric Functions

In the binomial expansion

$$(1 + x)^n = 1 + nx + n(n-1)\frac{x^2}{2!} + n(n-1)(n-2)\frac{x^3}{3!} + \cdots$$

we replace x by x/n to get

$$\left(1 + \frac{x}{n}\right)^n = 1 + n\frac{x}{n} + n(n-1)\frac{1}{2!}\left(\frac{x}{n}\right)^2$$

$$+ n(n-1)(n-2)\frac{1}{3!}\left(\frac{x}{n}\right)^3 + \cdots$$

$$= 1 + x + \frac{n(n-1)}{n^2}\frac{x^2}{2!} + \frac{n(n-1)(n-2)}{n^3}\frac{x^3}{3!} + \cdots.$$

If we now take the limit as $n \to \infty$ of both sides, the expressions involving n on the right-hand side of the expression approach 1 and we have

$$\lim_{n \to \infty} \left(1 + \frac{x}{n}\right)^n$$

$$= \lim_{x \to \infty} \left(1 + x + \frac{n(n-1)}{n^2}\frac{x^2}{2!} + \frac{n(n-1)(n-2)}{n^3}\frac{x^3}{3!} + \cdots\right)$$

$$= 1 + x + \frac{x^2}{2!} + \frac{x^3}{3!} + \cdots = e^x.$$

The Taylor expansions,

$$e^x = 1 + x + \frac{x^2}{2!} + \frac{x^3}{3!} + \cdots,$$

$$\sin x = x + \frac{x^3}{3!} + \frac{x^5}{5!} + \cdots,$$

$$\cos x = 1 + \frac{x^2}{2!} + \frac{x^4}{4!} + \cdots,$$

can be extended to complex numbers by replacing $x \in \mathbb{R}$ by $z \in \mathbb{C}$ and, in particular, by replacing x by ix. This results in

$$e^{ix} = 1 + (ix) + \frac{(ix)^2}{2!} + \frac{(ix)^3}{3!} + \cdots$$

$$= 1 + ix - \frac{x^2}{2!} - \frac{ix^3}{3!} + \cdots$$

$$= \left(1 - \frac{x^2}{2!} + \cdots\right) + i\left(x - \frac{x^3}{3!} + \cdots\right)$$

$$= \cos x + i \sin x.$$

Now replace ix by $-ix$ to get $e^{-ix} = \cos x - i \sin x$ and therefore

$$\sin x = \frac{1}{2i}(e^{ix} - e^{-ix}) \quad \text{and} \quad \cos x = \frac{1}{2}(e^{ix} + e^{-ix}).$$

$\log_{10} 2$ Is Irrational

Here we prove that $\log_{10} 2$ *is* irrational. In fact, $\log_{10} a$ is irrational for all a not a rational power of 10, but we will satisfy ourselves with the special case we need in the text.

The standard argument is to assume otherwise and so write

$$\log_{10} 2 = \frac{p}{q},$$

where p and q are integers.

This means that $10^{p/q} = 2$ and so $10^p = 2 \times 10^q$ and this means that $2^p \times 5^p = 2 \times 2^q \times 5^q = 2^{q+1} \times 5^q$, which means that $p = q+1$ and $p = q$, which is an obvious contradiction.

Floor and Ceiling Functions

The floor function $\lfloor x \rfloor$ is defined as the greatest integer not exceeding x and the ceiling function $\lceil x \rceil$ as the smallest integer greater than or equal to x.

For example:

$$\lfloor 8.15 \rfloor = 8,$$
$$\lceil 8.15 \rceil = 9,$$
$$\lfloor -8.15 \rfloor = -9,$$
$$\lceil -8.15 \rceil = -8,$$
$$\lfloor 8 \rfloor = 8,$$
$$\lceil 8 \rceil = 8.$$

Figure 4 shows the 'staircase' behaviour of each of the functions, with the rises occurring in intervals of length 1 on the horizontal axis.

We list several basic properties:

●

$$\lceil x \rceil = \begin{cases} x = \lfloor x \rfloor, & x \text{ an integer} \\ \lfloor x \rfloor + 1, & x \text{ not an integer.} \end{cases}$$

- If $\{x\}$ denotes the decimal part of the real number x, $\{x\} = x - \lfloor x \rfloor$,
- $\lfloor x + k \rfloor = \lfloor x \rfloor + k$, $\lceil x + k \rceil = \lceil x \rceil + k$, for any integer k,
- $x < y \Rightarrow \lfloor x \rfloor \leqslant \lfloor y \rfloor$ and $\lceil x \rceil \leqslant \lceil y \rceil$.

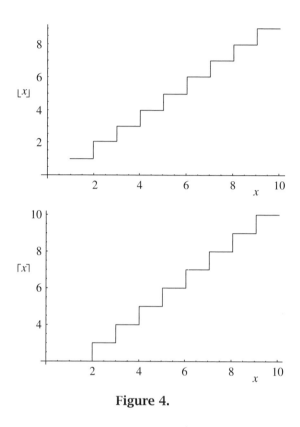

Figure 4.

The Pigeon-Hole Principle

Otherwise known as the Dirichlet drawer principle (named after Lejeune Dirichlet, who in 1834 mentioned it under the name Schubfachprinzip and not to be confused with the Dirichlet principle in potential theory). In modern times its familiar name readily suggests the imagery that is used to express the result. There are two forms, the first of which is:

> If $n + 1$ pigeons are distributed among n holes, then at least two of them occupy the same hole.

The obvious truth of the statement encourages the thought that it is no more than a trivial observation—but that is very far indeed from reality. We give just two examples of it being put to subtle use.

Prove that if seven distinct numbers are selected from $\{1, 2, 3, 4, 5, 6, 7, 8, 9, 10, 11\}$, then two of them will sum to 12.

The list $\{1, 11\}$, $\{2, 10\}$, $\{3, 9\}$, $\{4, 8\}$, $\{5, 7\}$, $\{6\}$ exhausts the possible ways of choosing pairs of the numbers which sum to 12; it also contains the single number, 6.

Here the 'holes' are these six subsets in the list and the 'pigeons' the seven chosen numbers: since we are choosing seven different numbers, at least two must come from the same subset containing a pair of numbers—and therefore sum to 12.

If each point of the plane is randomly coloured red or blue, then there exists a rectangle with all four vertices of the same colour.

We will find a rectangle whose sides are horizontal and vertical. First, draw any three horizontal lines. Any vertical line drawn intersects these three horizontal lines in three points and these may be coloured with two colours in $2^3 = 8$ different ways (the pigeon holes). Choose any nine vertical lines (the pigeons), then there are bound to be at least two triplets of points coloured in the same manner (for example, RRB and RRB): select two such triplets to define the vertical sides. In any triplet, at least two points are of the same colour. Select two such and we have the horizontal sides.

The more general form of the principle is:

If n or more pigeons are distributed among $k(< n)$ holes, then at least one hole contains at least $\lceil n/k \rceil$ pigeons.

To prove this, suppose that each pigeon hole contains at most $\lceil n/k \rceil - 1$ pigeons. Then the total number of pigeons is at most

$$
k\left(\left\lceil \frac{n}{k} \right\rceil - 1\right) = \begin{cases} k\left(\dfrac{n}{k} - 1\right), & \dfrac{n}{k} \text{ an integer,} \\[2ex] k\left\lfloor \dfrac{n}{k} \right\rfloor, & \dfrac{n}{k} \text{ not an integer.} \end{cases}
$$

In either case, this is strictly less than n and we have a contradiction.

Two examples of the use of this follow:

Suppose there are 49 people in a room, then at least five of them must have their birthday in the same month.

Here, the 49 pigeons are to be placed in 12 holes (one for each month). Using the above principle, there must be at least $\lceil \frac{49}{12} \rceil = 5$ pigeons in the same hole—or birthdays in the same month.

How many cards must be selected from a pack of 52 to guarantee that at least three cards of the same suit are chosen?

Here, there are n pigeons to fit into four holes (one for each suit). We require least n so that $\lceil n/4 \rceil \geqslant 3$. A small exercise in arithmetic reveals that this is $n = 9$.

Logs and Floors

In chapter 15 we needed that, if $M \geqslant 1$ is a positive integer,

$$\lfloor \log_{10}(M + 1) \rfloor = \begin{cases} \lfloor \log_{10} M \rfloor, & M + 1 \text{ not a power of } 10, \\ \lfloor \log_{10} M \rfloor + 1, & M + 1 \text{ a power of } 10. \end{cases}$$

This can be proved as follows:

$$0 < \log_{10} \left(1 + \frac{1}{M} \right) < 1$$

and so

$$0 < \log_{10} \left(\frac{M + 1}{M} \right) < 1,$$
$$0 < \log_{10}(M + 1) - \log_{10} M < 1,$$
$$\log_{10} M < \log_{10}(M + 1) < \log_{10} M + 1.$$

Using the properties of the floor function,

$$\lfloor \log_{10} M \rfloor \leqslant \lfloor \log_{10}(M + 1) \rfloor \leqslant \lfloor \log_{10} M + 1 \rfloor,$$
$$\lfloor \log_{10} M \rfloor \leqslant \lfloor \log_{10}(M + 1) \rfloor \leqslant \lfloor \log_{10} M \rfloor + 1.$$

This means that the middle integer is sandwiched between the two consecutive outside integers and this can only be reconciled by one or other of the two alternatives:

$$\lfloor \log_{10}(M + 1) \rfloor = \lfloor \log_{10} M \rfloor \quad \text{or} \quad \lfloor \log_{10}(M + 1) \rfloor = \lfloor \log_{10} M \rfloor + 1.$$

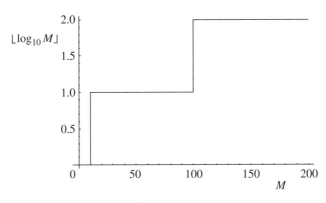

Figure 5.

With $\lfloor x \rfloor$ the rises appear in intervals of 1 unit; the $\log_{10}$ function changes this to intervals that are integer powers of 10, as indicated in figure 5. From these facts the result follows.

A Rational Approximation to an Irrational Number

Here we will prove the approximation theorem of a rational number to an irrational number which was used in chapter 15. Its statement is:

> Given any irrational number λ and any positive integer k, there is a rational number m/n with $n \leqslant k$ such that $0 < \lambda - m/n < 1/nk$.

The proof is as follows.

For the given k consider the set of k irrational numbers $\{n\lambda : n = 1, 2, 3, \ldots, k\}$ and write $\alpha_n = \lfloor n\lambda \rfloor$ and $\beta_n = \{n\lambda\}$ for $n = 1, 2, 3, \ldots, k$. This means that $0 < \beta_n < 1$ and, since $n\lambda = \alpha_n + \beta_n$, $\beta_n = n\lambda - \alpha_n$.

Now divide the unit interval into k equal subintervals $I_1, I_2, I_3, \ldots, I_k$, as shown in figure 4. Evidently each of the β_n is irrational and so cannot be any of the subinterval endpoints

$$0, \frac{1}{k}, \frac{2}{k}, \frac{3}{k}, \ldots, \frac{k}{k},$$

which means that they lie strictly within the intervals.

Now take two cases.

At least one of the β_n lies in I_1. For that β_n it must be that $0 < \beta_n < 1/k$ and so $0 < n\lambda - \alpha_n < 1/k$ and

$$0 < \lambda - \frac{\alpha_n}{n} < \frac{1}{nk}.$$

Take $\alpha_n = m$ and we have what is required.

Now suppose that none of the β_n lie in I_1.

This means that all k of them lie in the $k - 1$ subintervals $I_2, I_3, \ldots, I_k$ and so, by the pigeon-hole principle, at least two of them must lie in the same subinterval: let us write two such as β_p and β_q, with $\beta_p > \beta_q$. Since both of these numbers lie inside an interval of length $1/k$, their positive difference must satisfy $0 < \beta_p - \beta_q < 1/k$ and so

$$0 < (p\lambda - \alpha_p) - (q\lambda - \alpha_q) < \frac{1}{k}$$

and this means that

$$0 < (p - q)\lambda - (\alpha_p + \alpha_q) < \frac{1}{k}$$

and

$$0 < \lambda - \frac{\alpha_p + \alpha_q}{p - q} < \frac{1}{k(p - q)}.$$

Take $\alpha_p + \alpha_q = m$ and $p - q = n$ and once more we have what we require.

This is not the tightest bound for such an approximation, but it is the one we want!

Index